WORK IN PROGRESS

A DECADE OF SKETCHES BY ILAN WAISBROD

studio GAIA

스튜디오 가이아, 상상의 공간 스케치북

초판 1쇄 2016년 3월 18일

글 · 그림: Ilan Waisbrod
발행인: 김기현
발행처: 시공문화사(**Spacetime**)
출판등록: 1993년 3월 12일
주소: 서울시 서대문구 독립문공원길 13 (03733) 극동프라자 5층
전화: 02) 3147-1212
팩스: 02) 3147-2626

http://www.spacetime.co.kr
mail: spacetime@korea.com

편집: Design Battery
인쇄: (주)예림인쇄사
제본: (주)다인바인텍
용지: (주)대림지업사

ISBN: 978-89-5592-345-2

정가: 25,000원

INTRODUCTION

Sketching and doodling has been a part of my life since my childhood. Just like any kid who enjoys drawing, I started copying and depicting shapes I saw around me at a very young age. I've always had a boundless imagination and I spent a lot of time using my hands as a tool to invent forms and bring ideas to life.

I loved creating new characters.

I drew two-dimensional and three-dimensional objects.

It was only later in life that I have become fascinated with architecture and design.

During my professional studies while I was taking industrial and interior design classes, I realized that being a designer is a very personal matter. Each designer needs to develop his/her own unique style.

Today, just like then, I'm constantly inspired to draw. Drawing has always been an obvious and essential tool for me to express my ideas. Only then, when an idea is brought to life, I started creating more practical and usable objects or environments. I follow my own intuition to create projects as a vehicle for change.

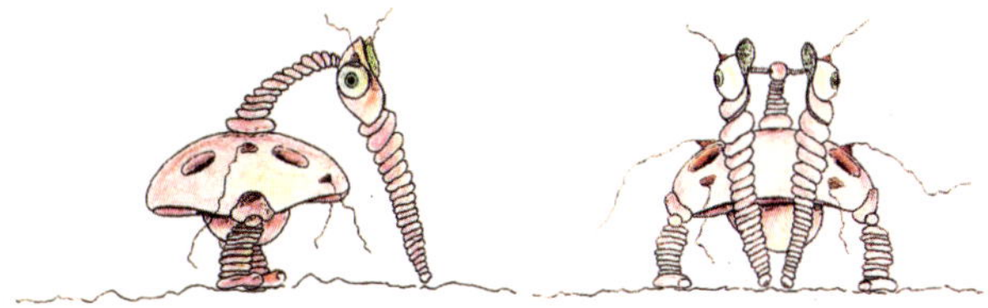

Guided by a holistic concept, I strive to find an appropriate design solution for my client that is practical yet transformational, innovative, structural, and emotional. I test my ideas as a medium to achieve value, inspire its user, and bring connections, while preserving my unique sense of style.

Ilan Waisbrod

It is my deep belief that design is an act of invention. Its real task begins once functional, emotional, and transformative needs are satisfied. It is not a hunger, but love and fear, and sometimes a wonder which make us create. Our aesthetics may change over time but this task remains the same.

Ilan Waisbrod

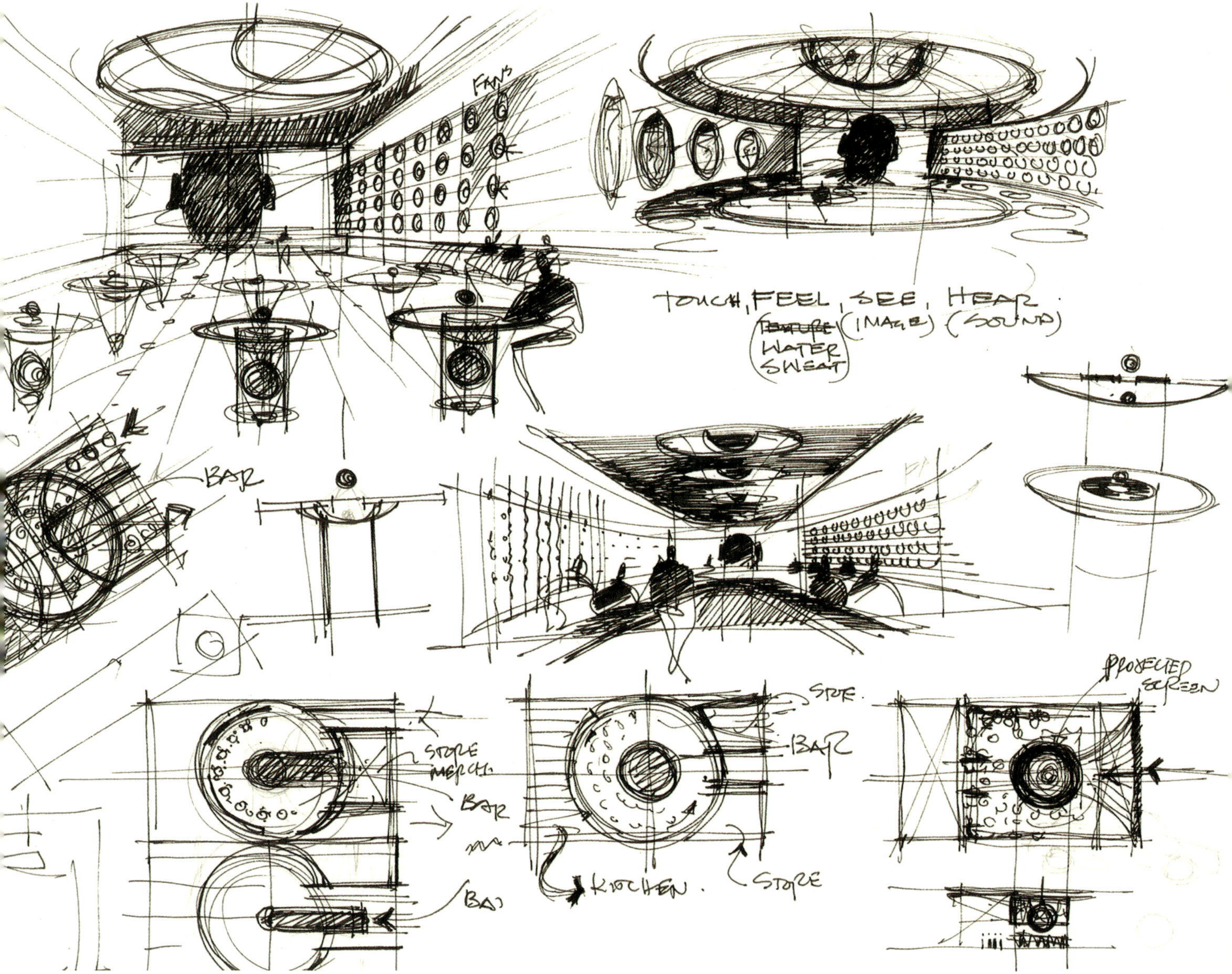
FANS
TOUCH, FEEL, SEE, HEAR.
(TEXTURE WATER SWEAT) (IMAGE) (SOUND)
BAR
STORE MERCH.
BAR
BAR
STORE.
BAR
KITCHEN.
STORE
PROJECTED SCREEN

VIEW TO KITCHEN.
8"x8" GLASS BLOCKS.
8 x 10 = 80"
SECTION - C - 1/4"=1'-0" SCALE
BANQ. PARTY ROOM.
ENTRANCE TO KITCEN.
23

AIR
23
JUMP

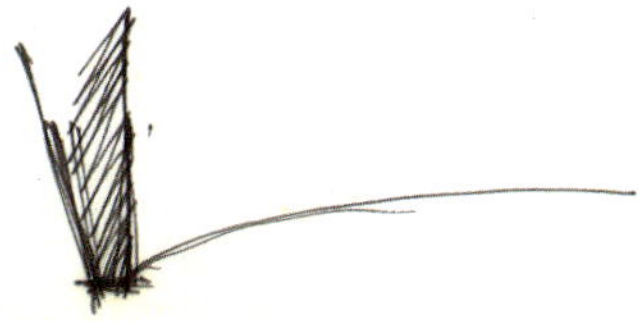

SECTION D – 1/4" SCALE

SECTION THROUGH CENTER SEATING AREA (B).

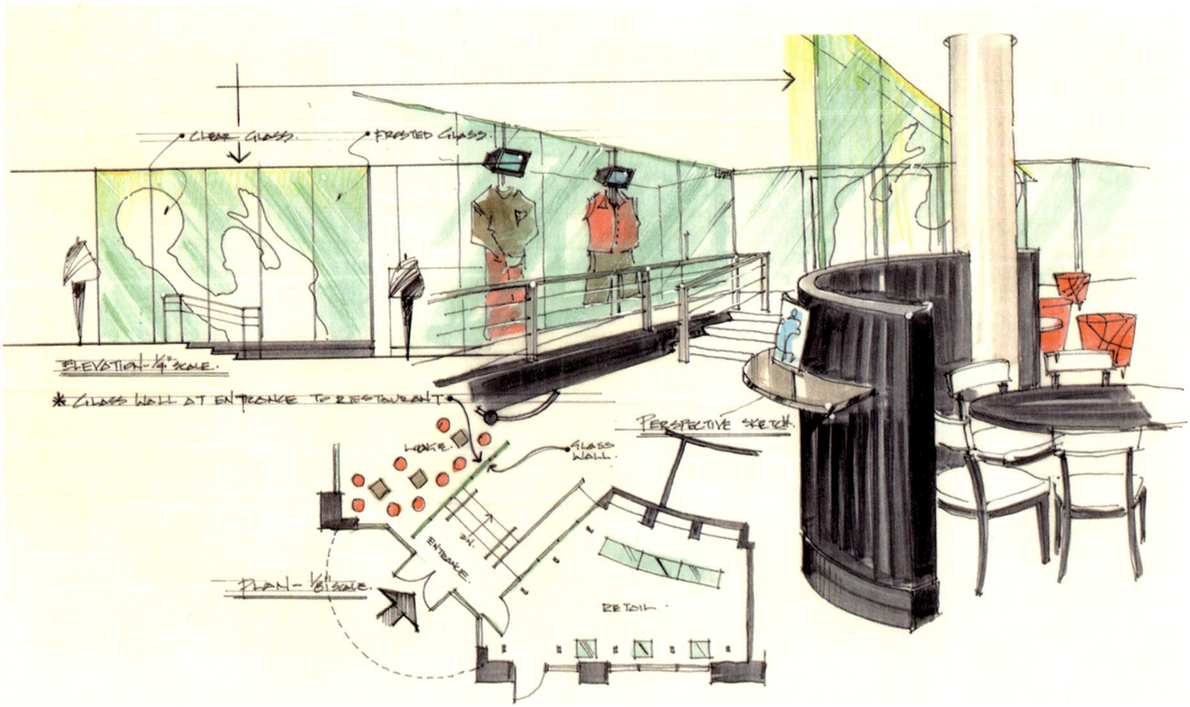
CLEAR GLASS.
FROSTED GLASS
ELEVATION-1/4" SCALE.
* GLASS WALL AT ENTRANCE TO RESTAURANT
PERSPECTIVE SKETCH
LOUNGE.
GLASS WALL.
ENTRANCE
DN.
PLAN-1/8" SCALE.
RETAIL

23
FLY
Real

JUMP
23
HIGHER
STOOLS..

MICHAEL JO
M.J

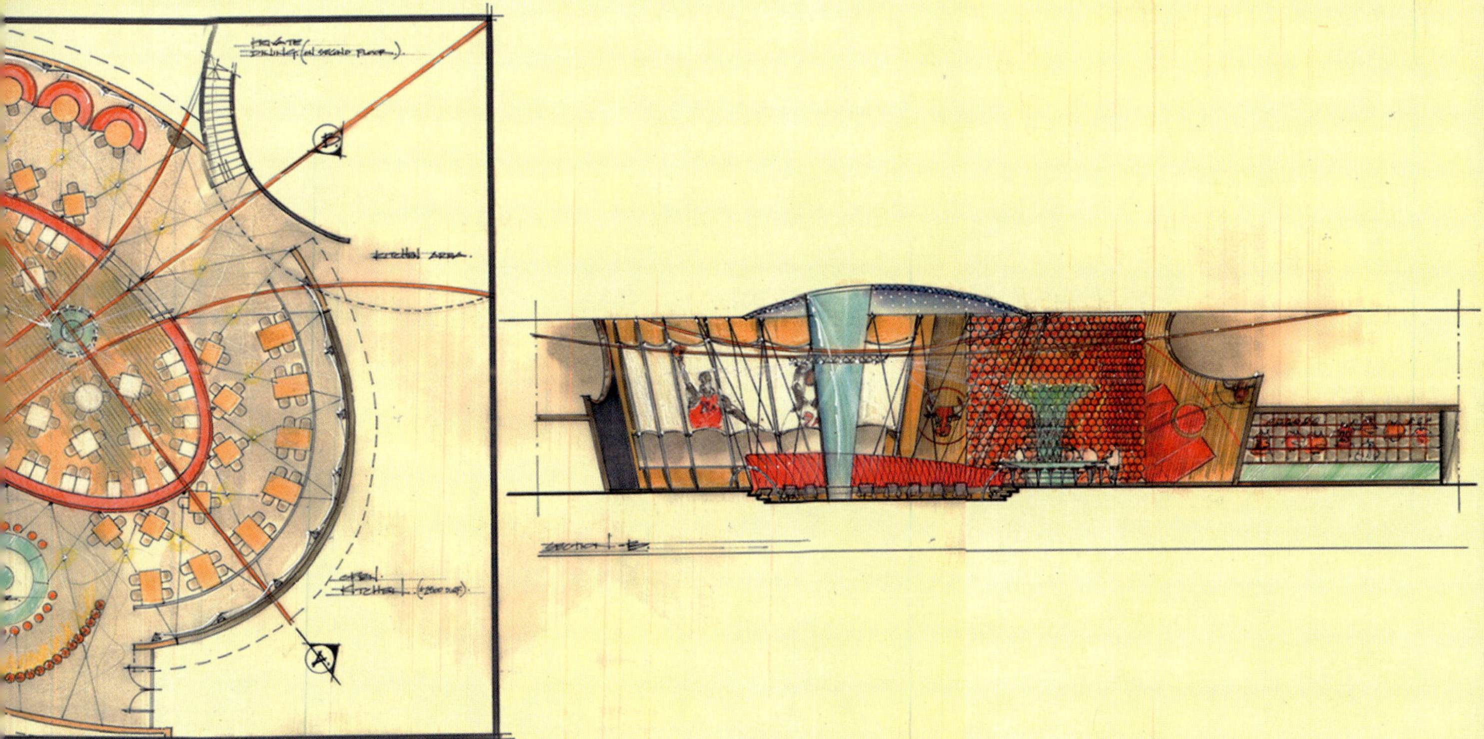
PRIVATE DINING (on second floor)
KITCHEN AREA

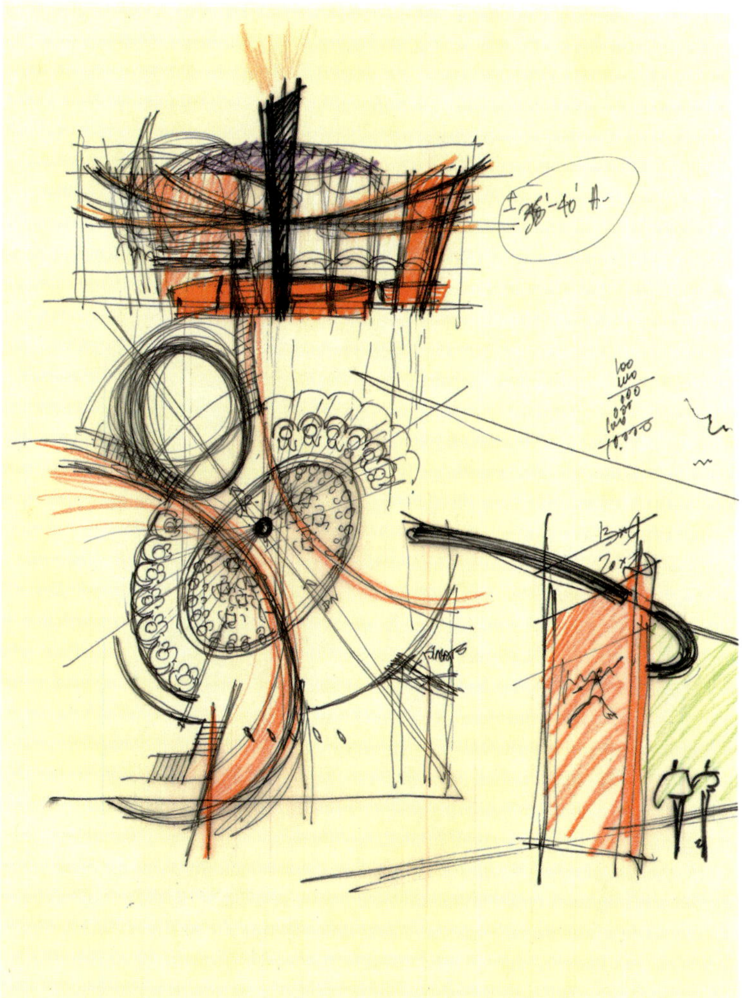

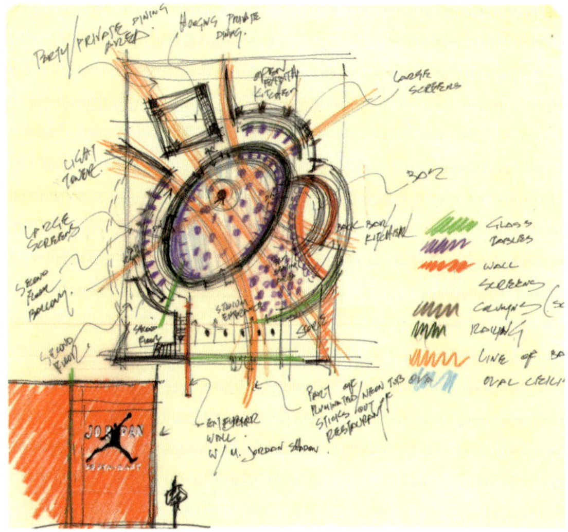
LARGE SCREENS
LIGHT TOWER
LARGE SCREENS

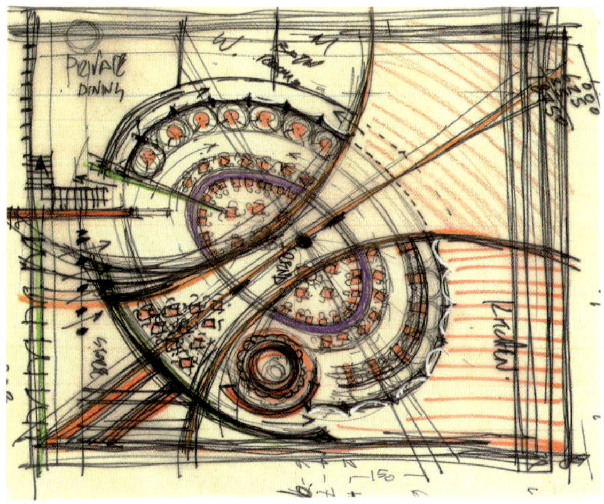
PRIVATE DINING
KITCHEN

23
Gatorade

NIKE

STONE
CUBES.

夜會
樓梯
電梯
餐具

STUDIO
GAIA
11 Garden Court Tenafly, N.J 07670
Tel: 201 541-1887 Fax 201 541-1887

STUDIO
GAIA
11 Garden Court Tenafly, N.J 07670
Tel: 201 541-1887 Fax 201 541-1887

STUDIO
GAIA
11 Garden Court Tenafly, N.J 07670
Tel: 201 541-1887 Fax 201 541-1887

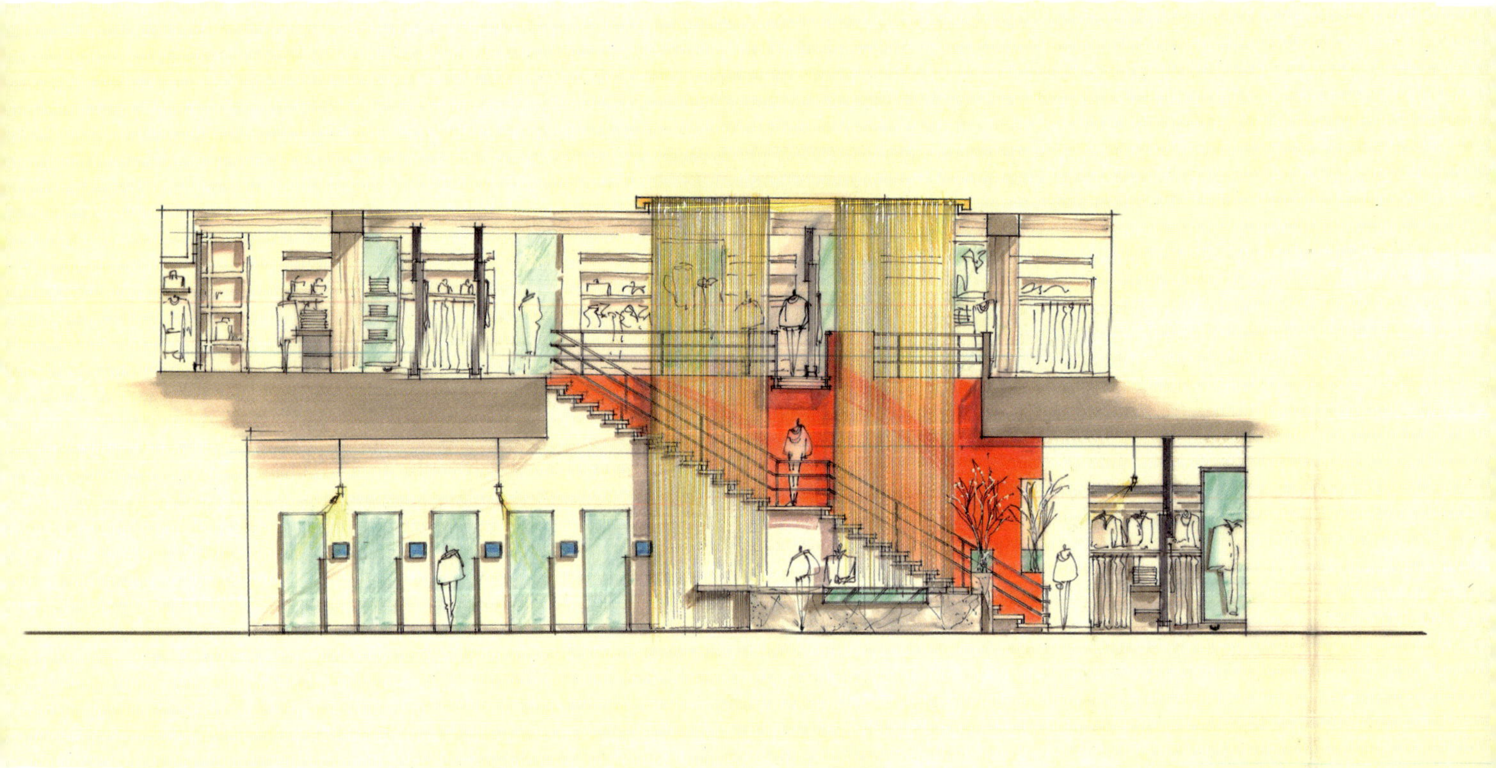

STUDIO
GAIA
11 Garden Court Tenafly, N.J 07670
Tel: 201 541-1887 Fax 201 541-1887

STUDIO
GAIA
11 Garden Court Tenafly, N.J 07670
Tel: 201 541-1887 Fax 201 541-1887

REPUBLIC

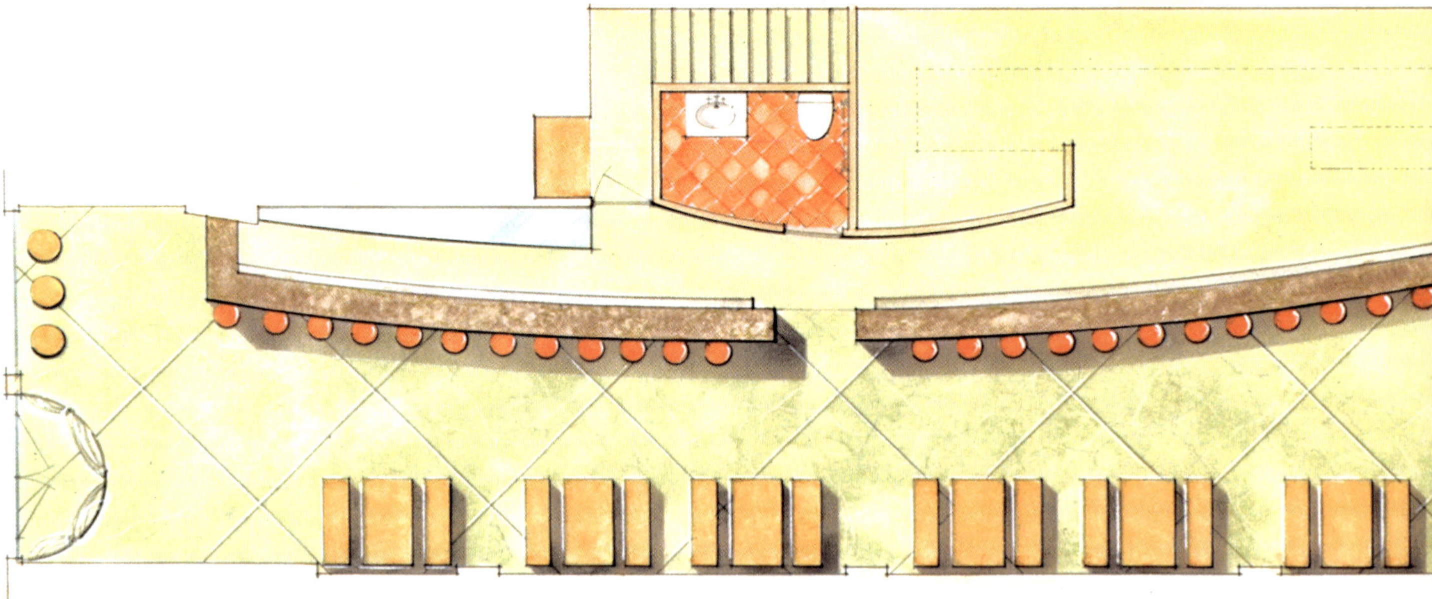

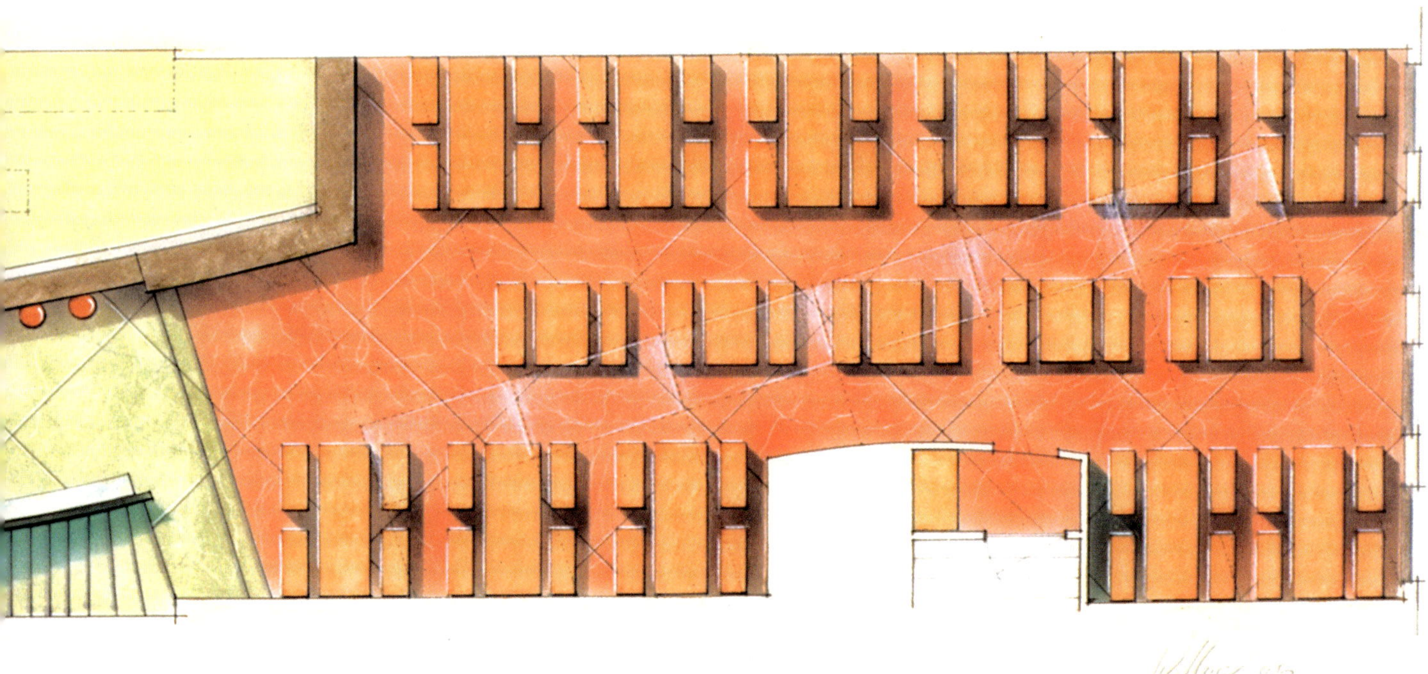

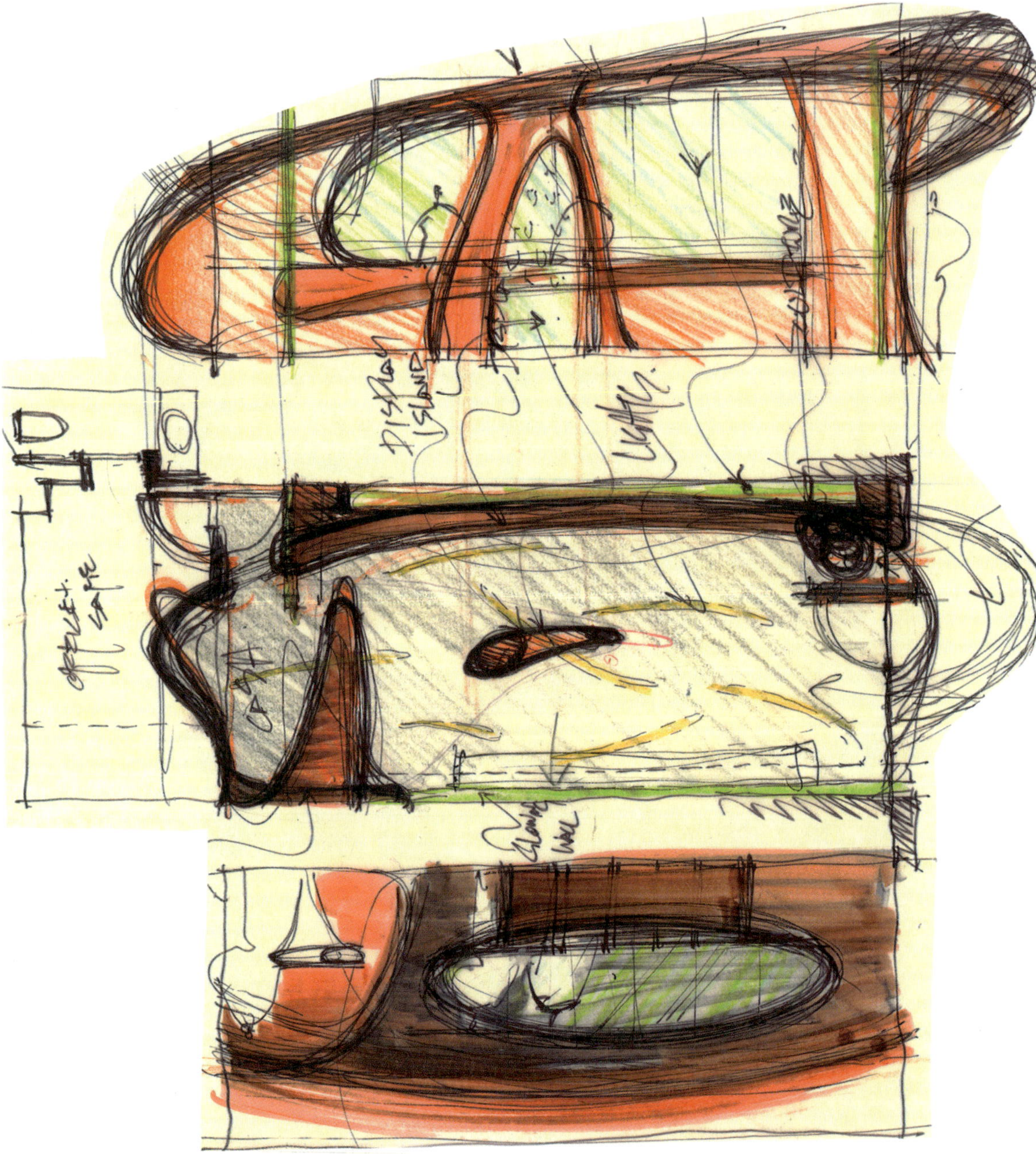
DISPLAY ISLAND
GLOWING WALL

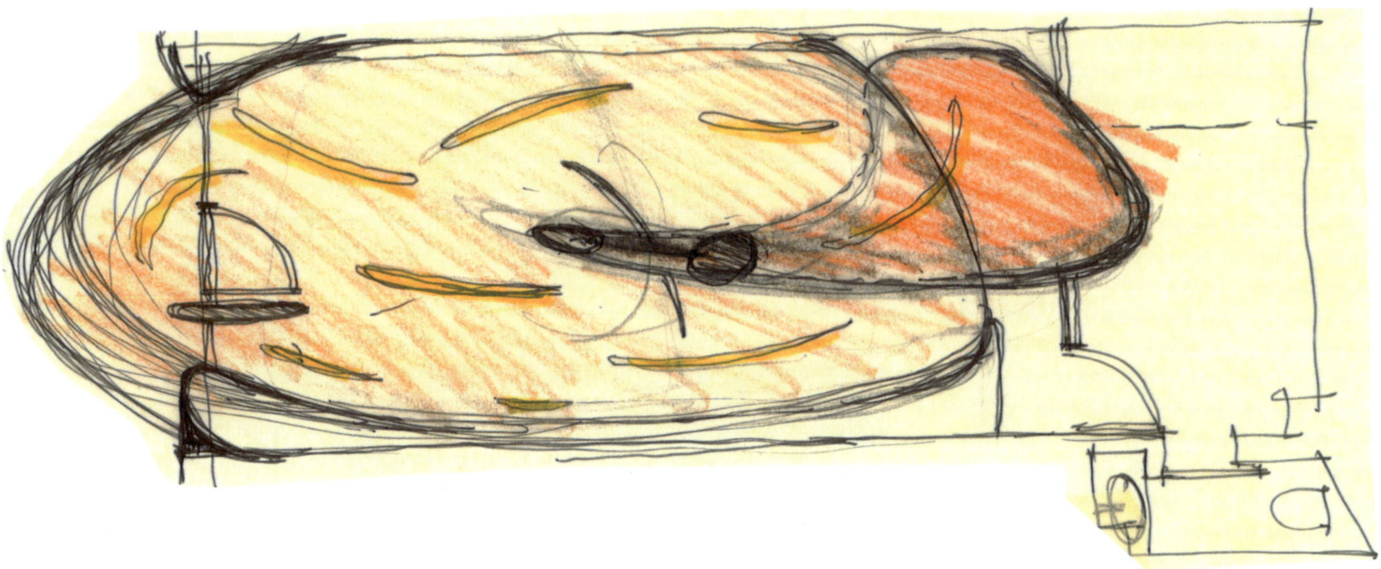

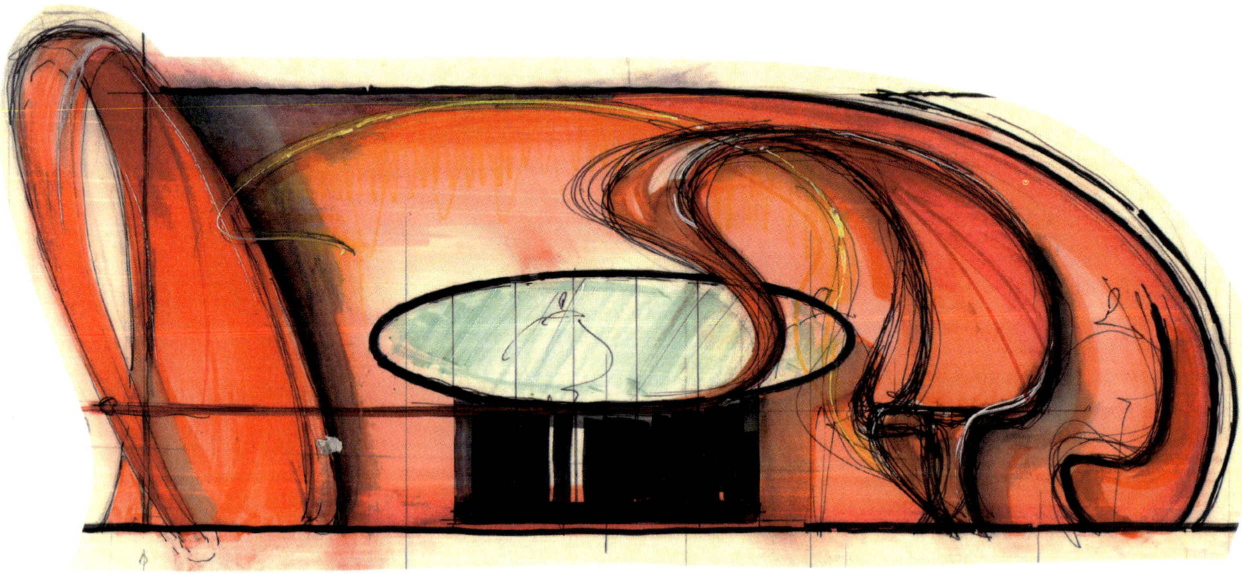

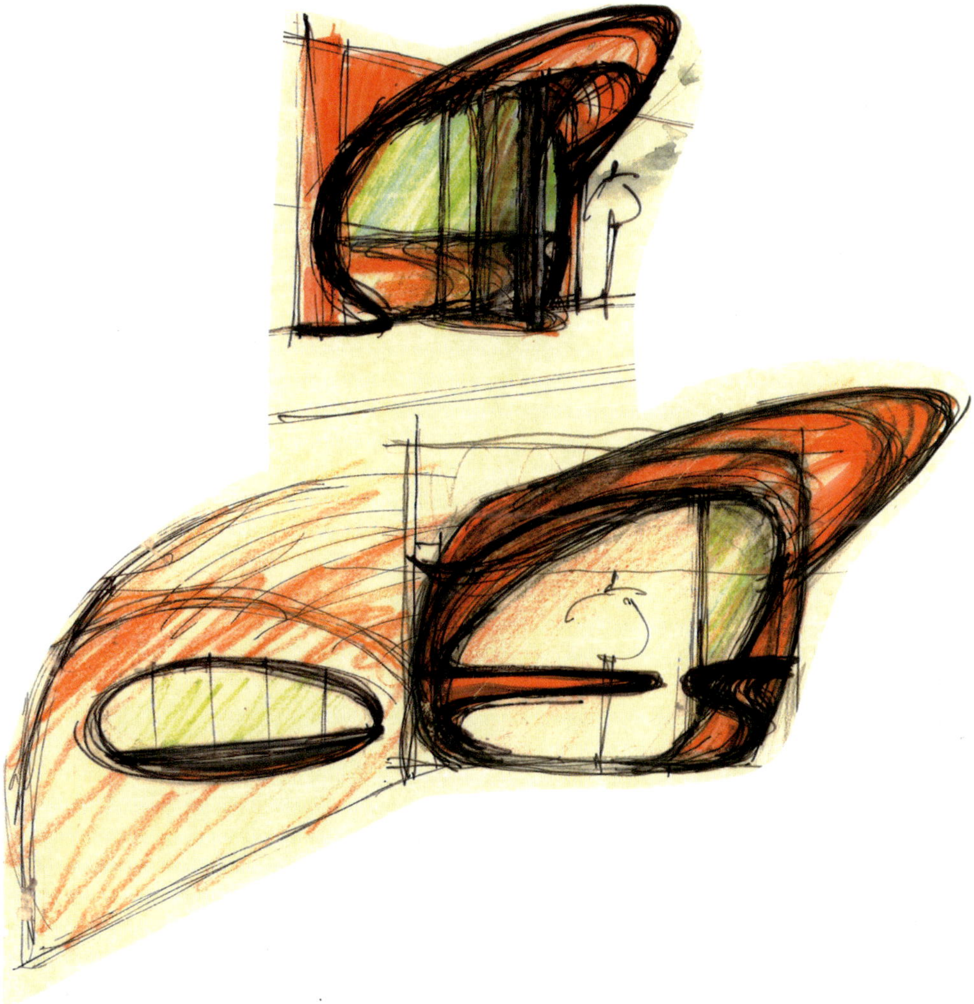

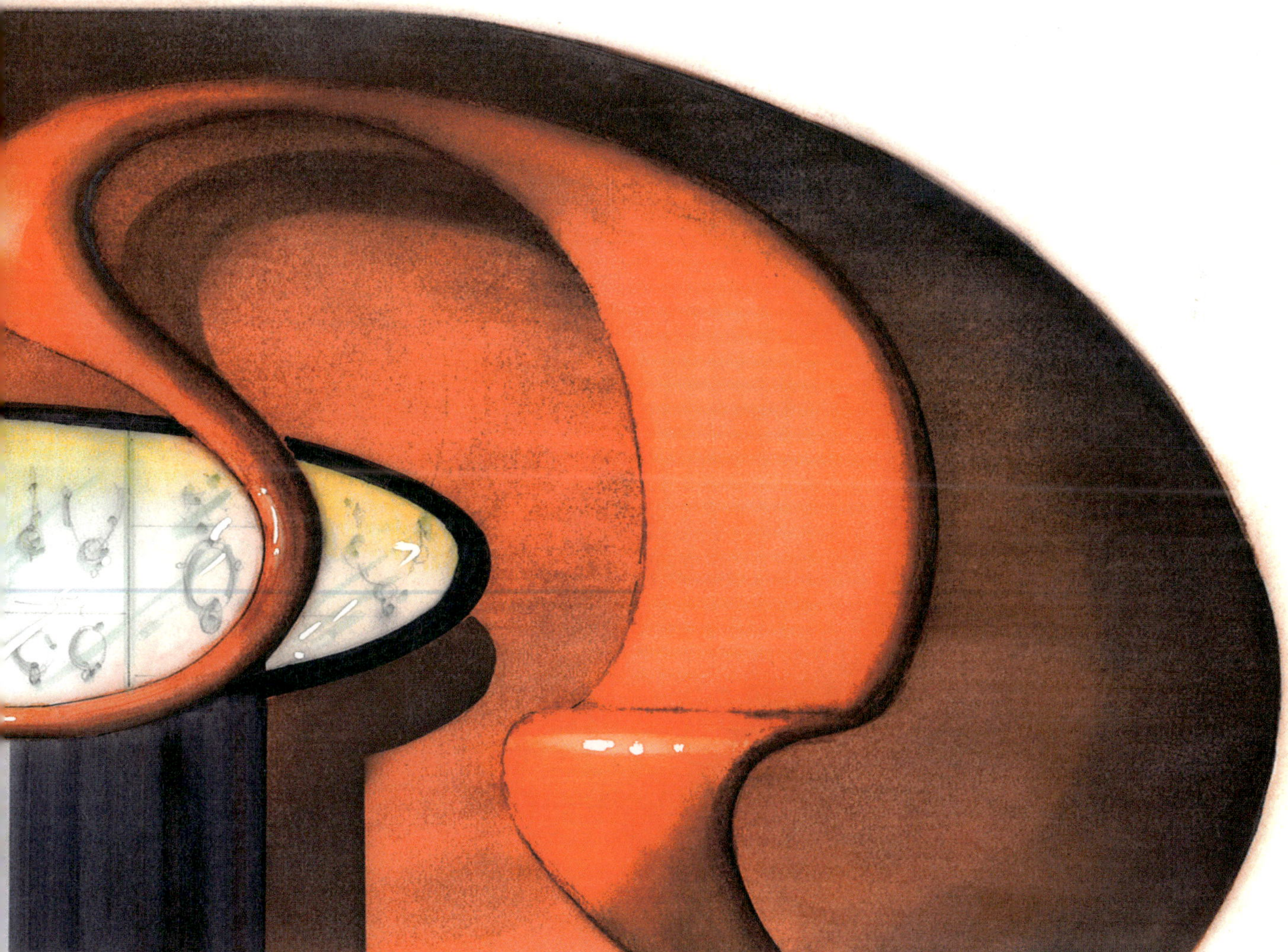

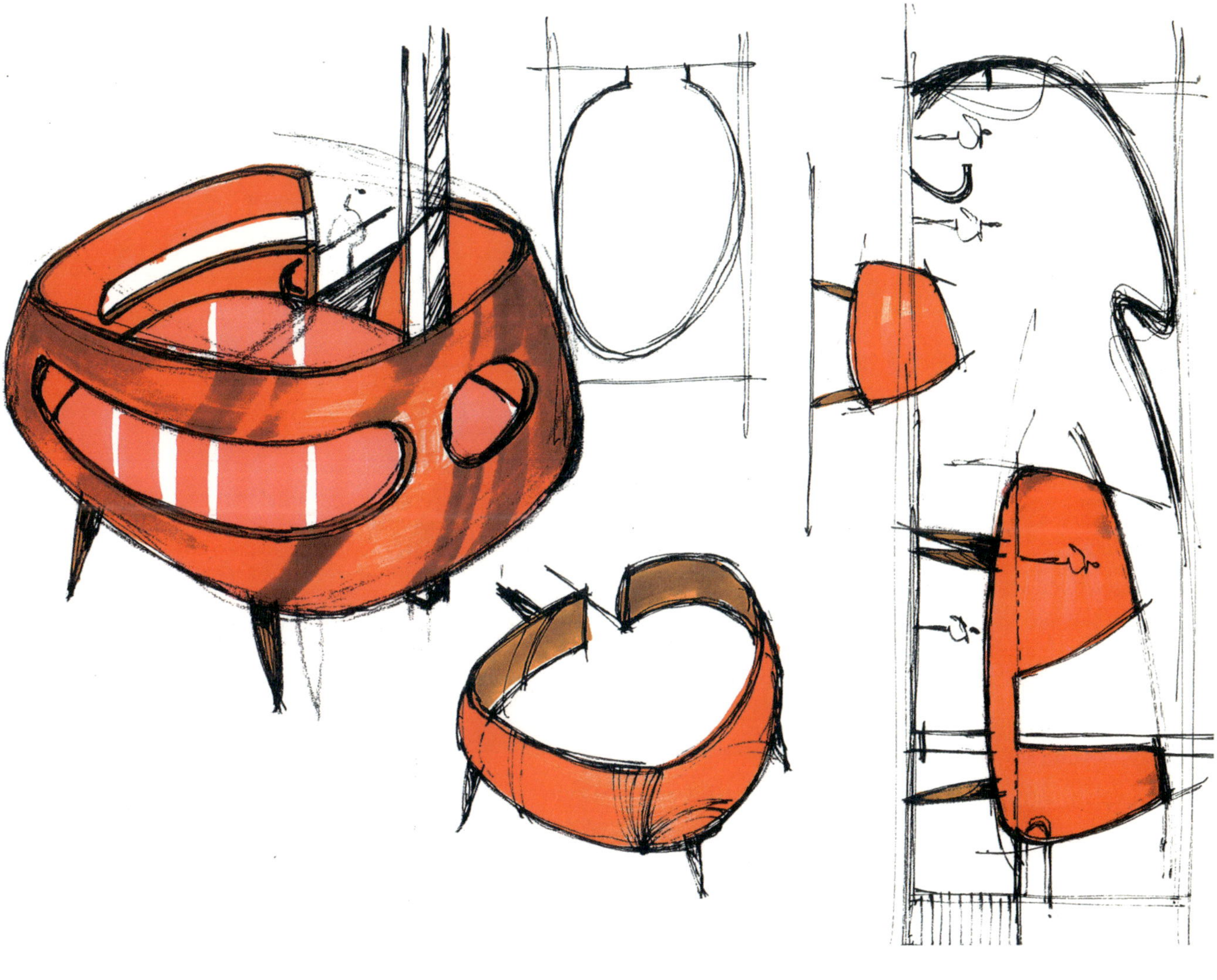

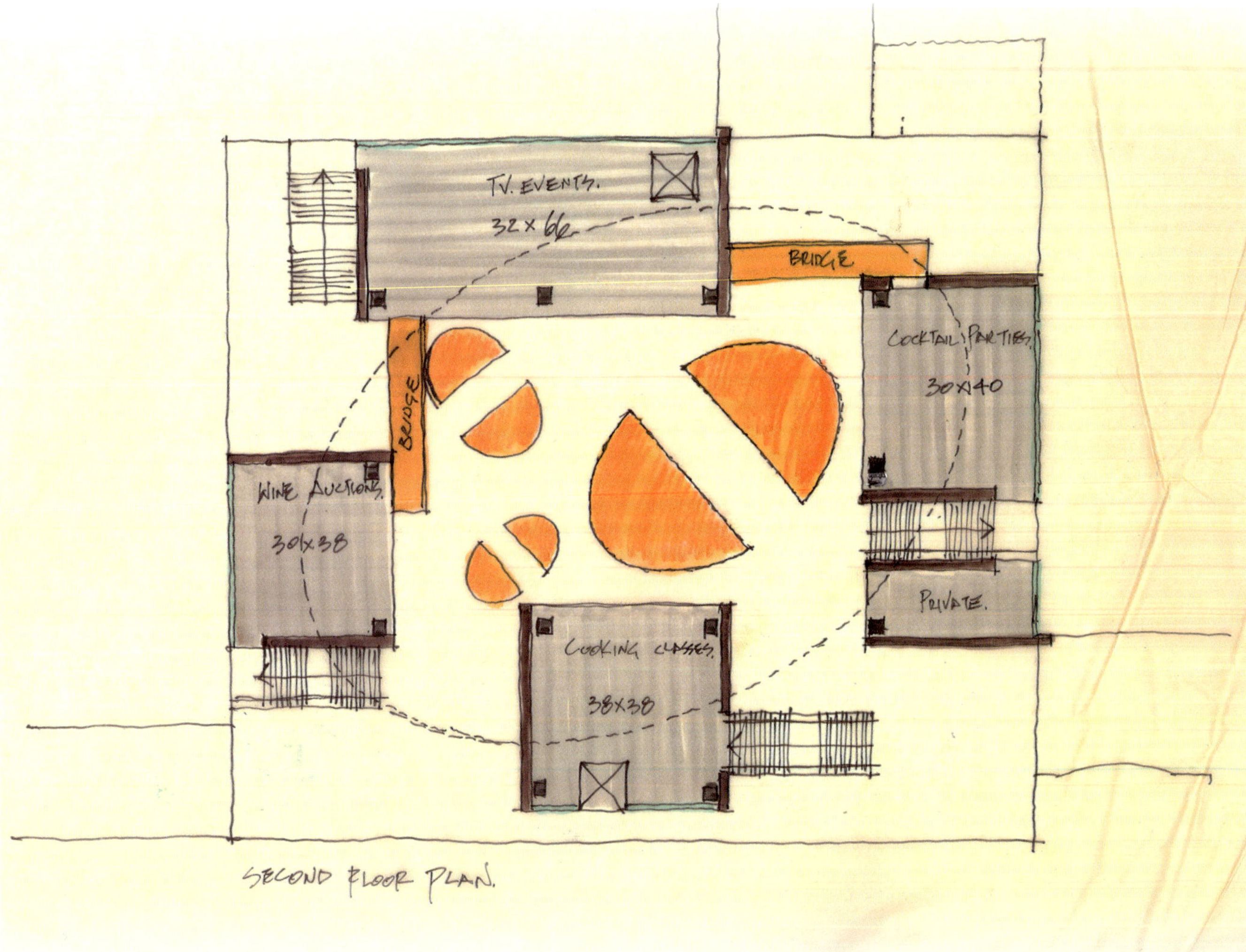
T.V. EVENTS.
32 x 66
BRIDGE
COCKTAIL PARTIES.
30x140
BRIDGE
WINE AUCTIONS.
30x38
PRIVATE.
COOKING CLASSES.
38x38
SECOND FLOOR PLAN.

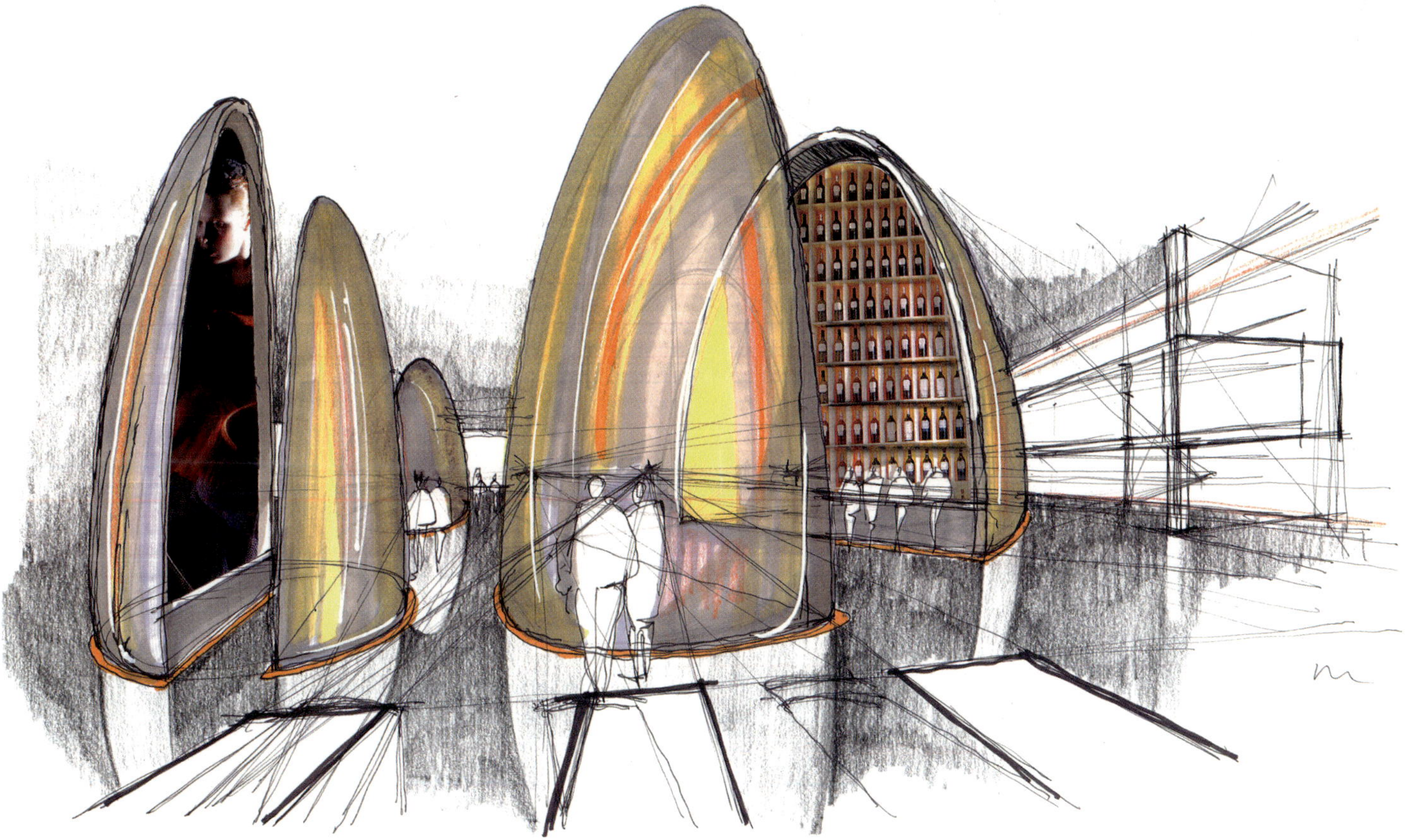

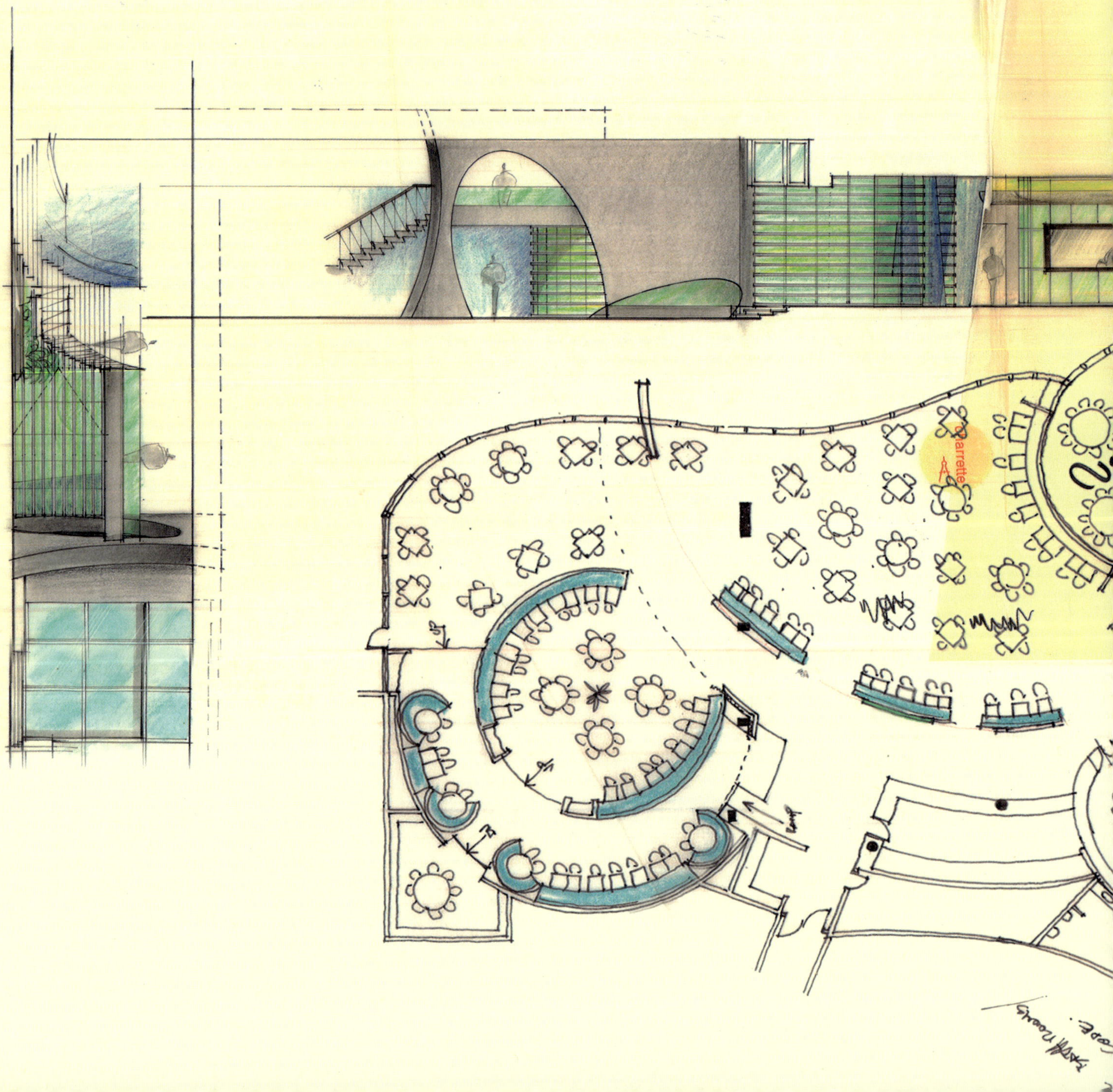

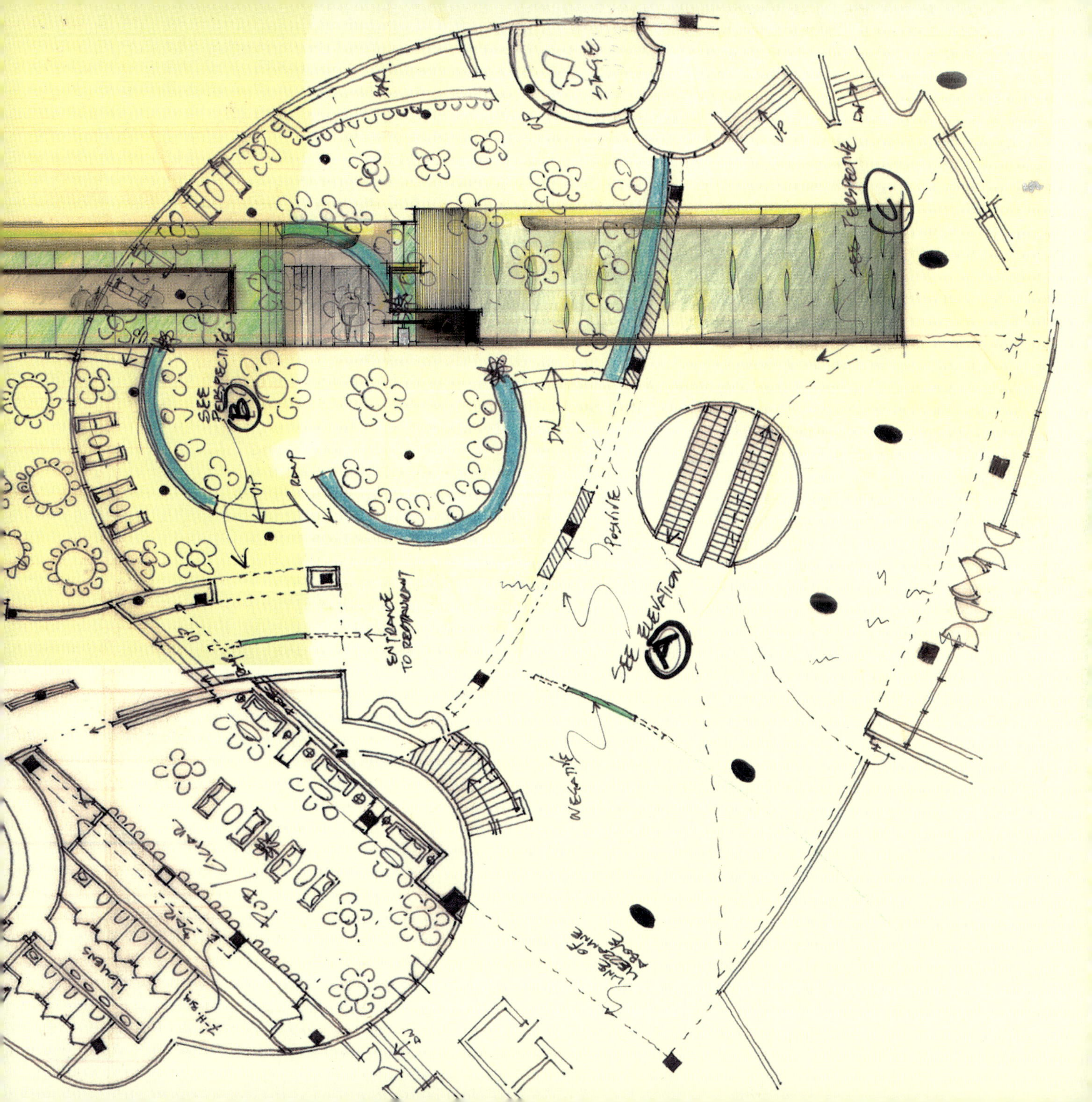
STAGE
BAR
UP
DN
SEE PERSPECTIVE C
SEE PERSPECTIVE B
RAMP
ENTRANCE TO RESTAURANT
SEE ELEVATION A
POSITIVE
NEGATIVE
WOMENS

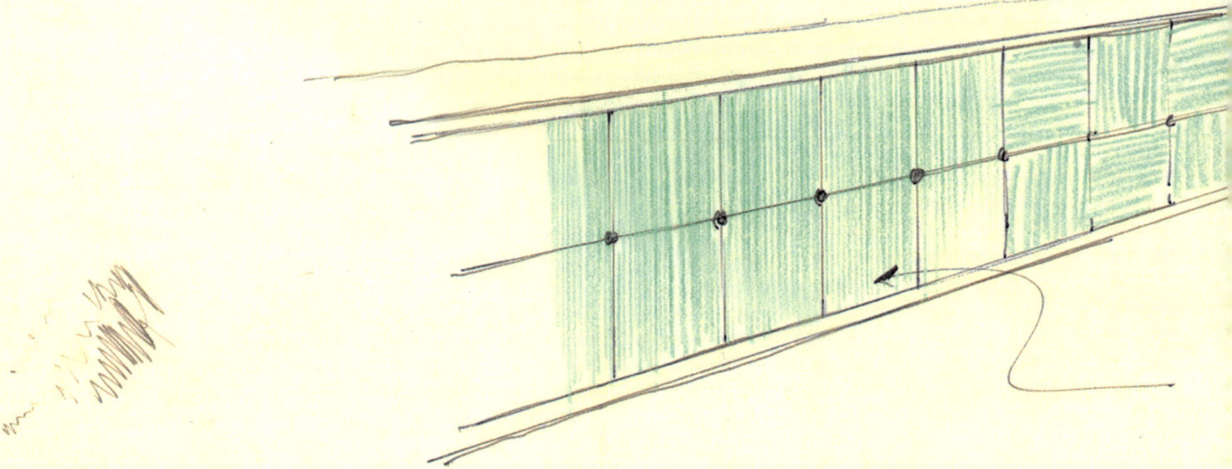

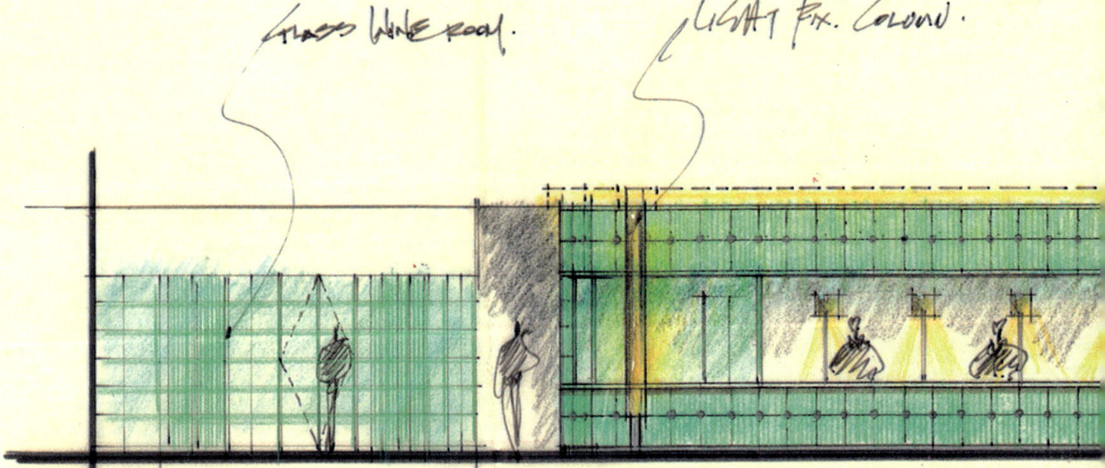

VIEW OF OPEN KITCHEN.

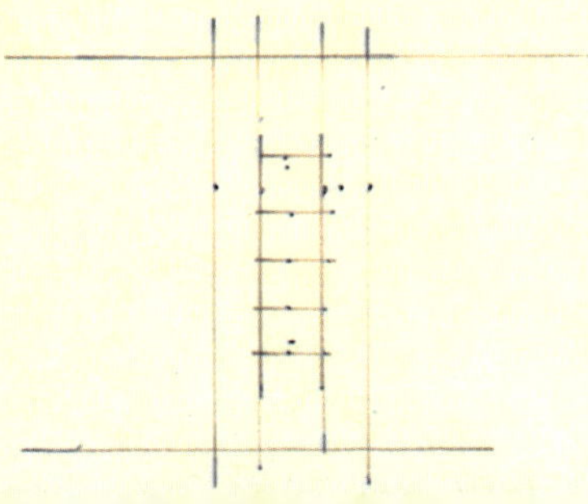

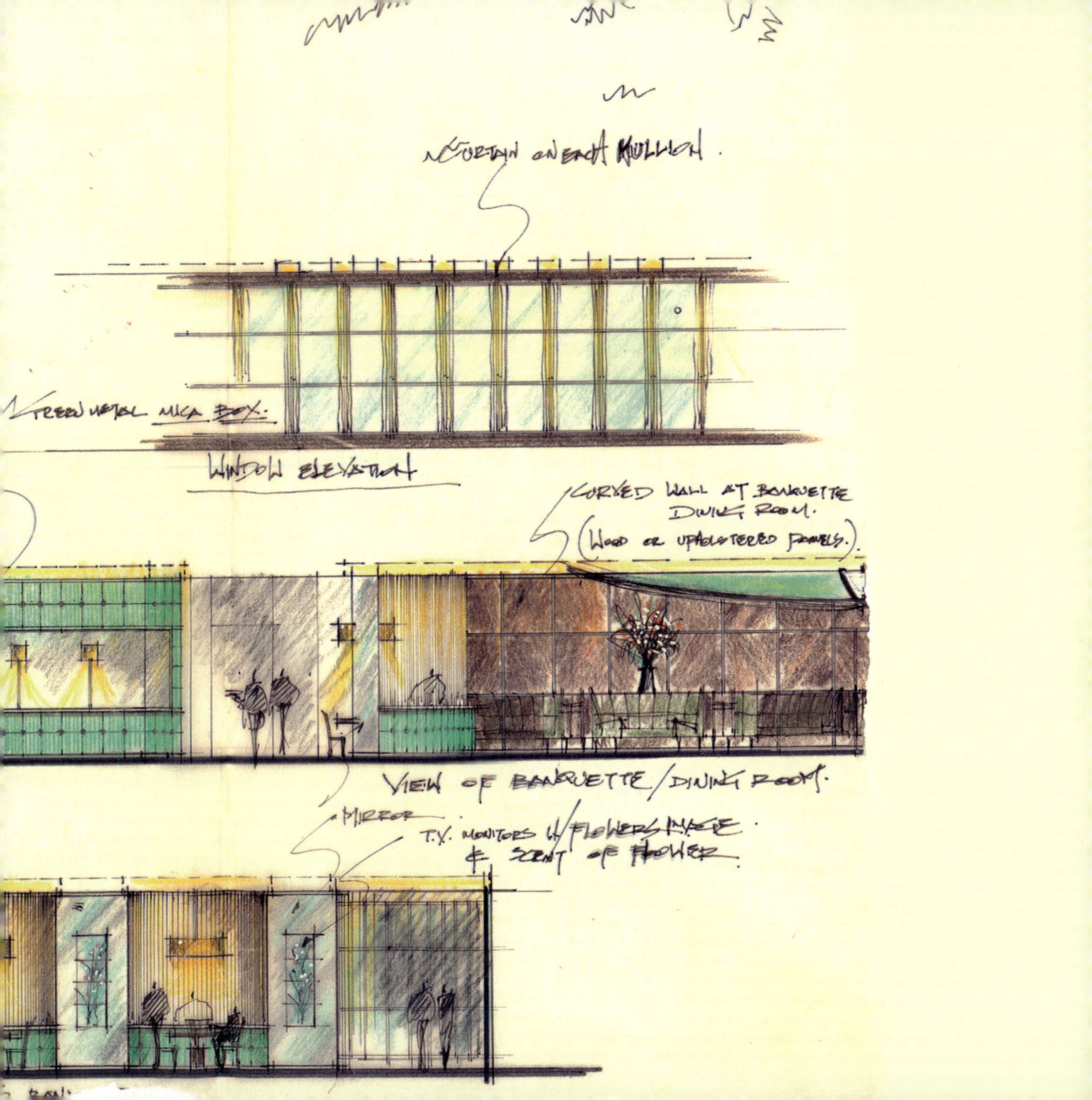
CURTAIN ON EACH MULLION.
GREEN METAL MICA BOX.
WINDOW ELEVATION
CURVED WALL AT BANQUETTE
DINING ROOM.
(WOOD OR UPHOLSTERED PANELS).
VIEW OF BANQUETTE / DINING ROOM.
MIRROR
T.V. MONITORS W/ FLOWERS IMAGE.
& SCENT OF FLOWER.

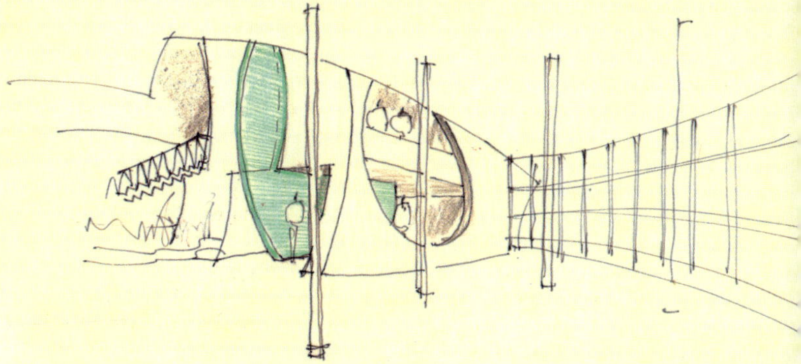

PERSPECTIVE (B)

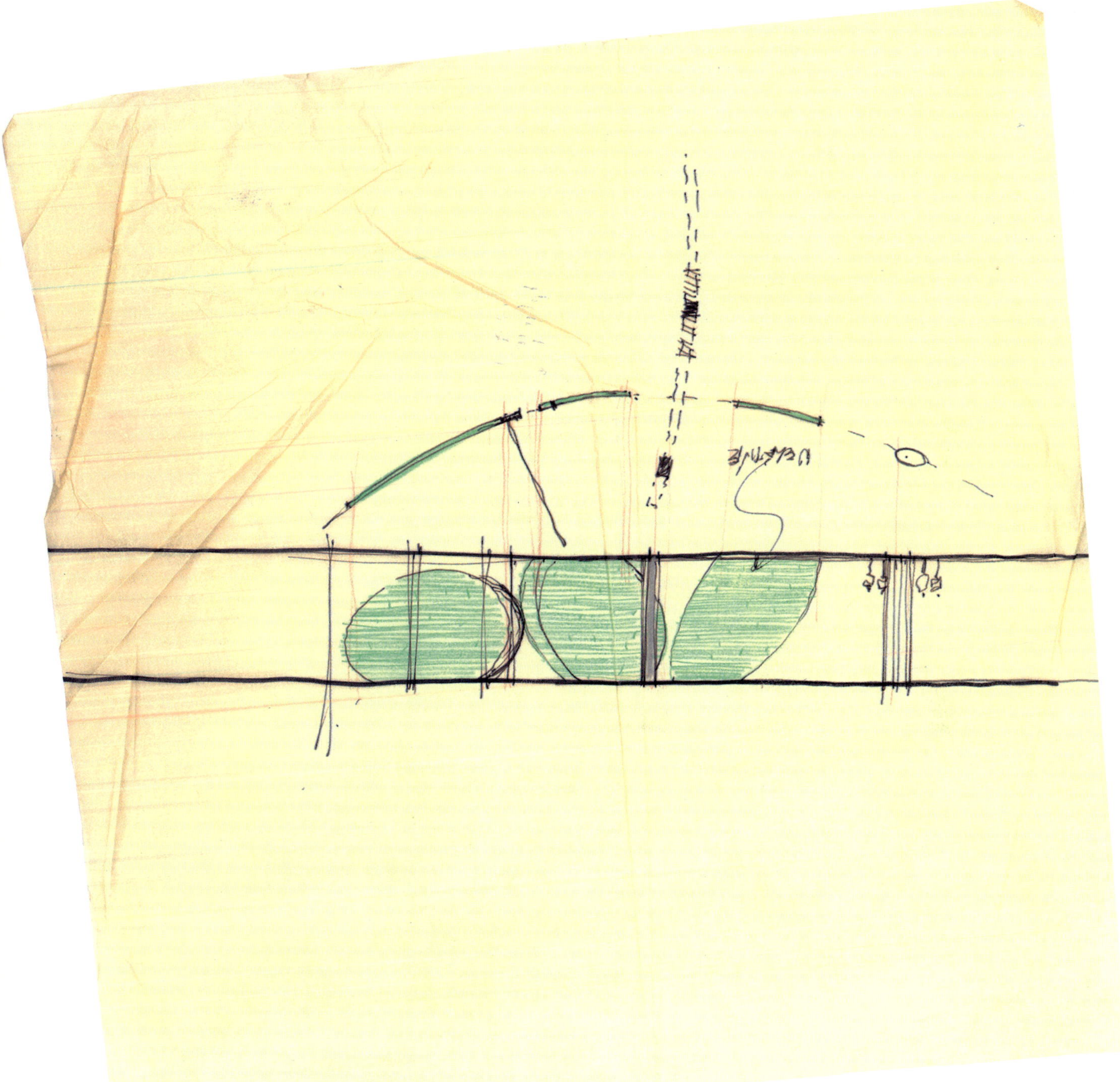

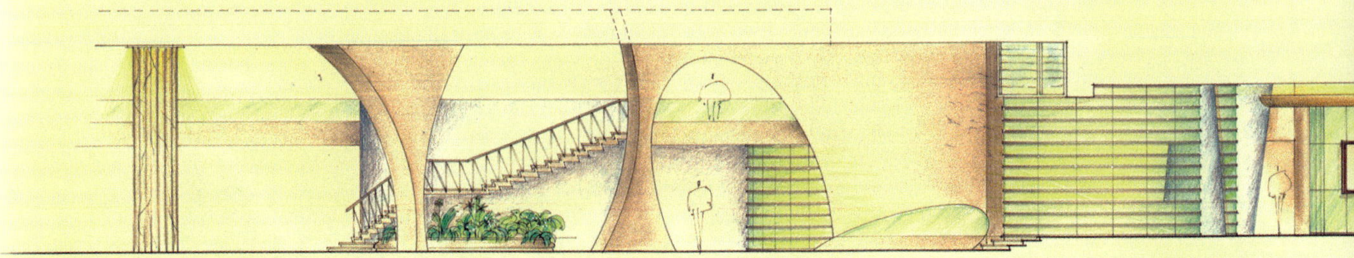

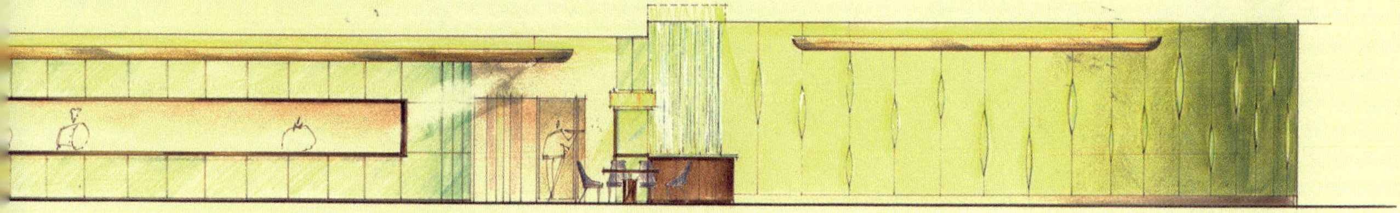

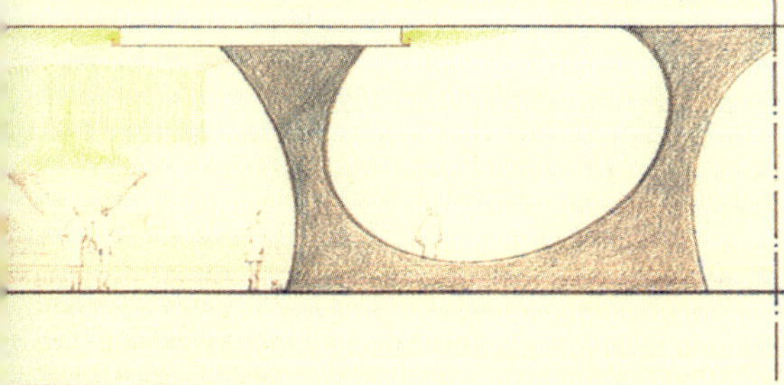

A

B

C

D

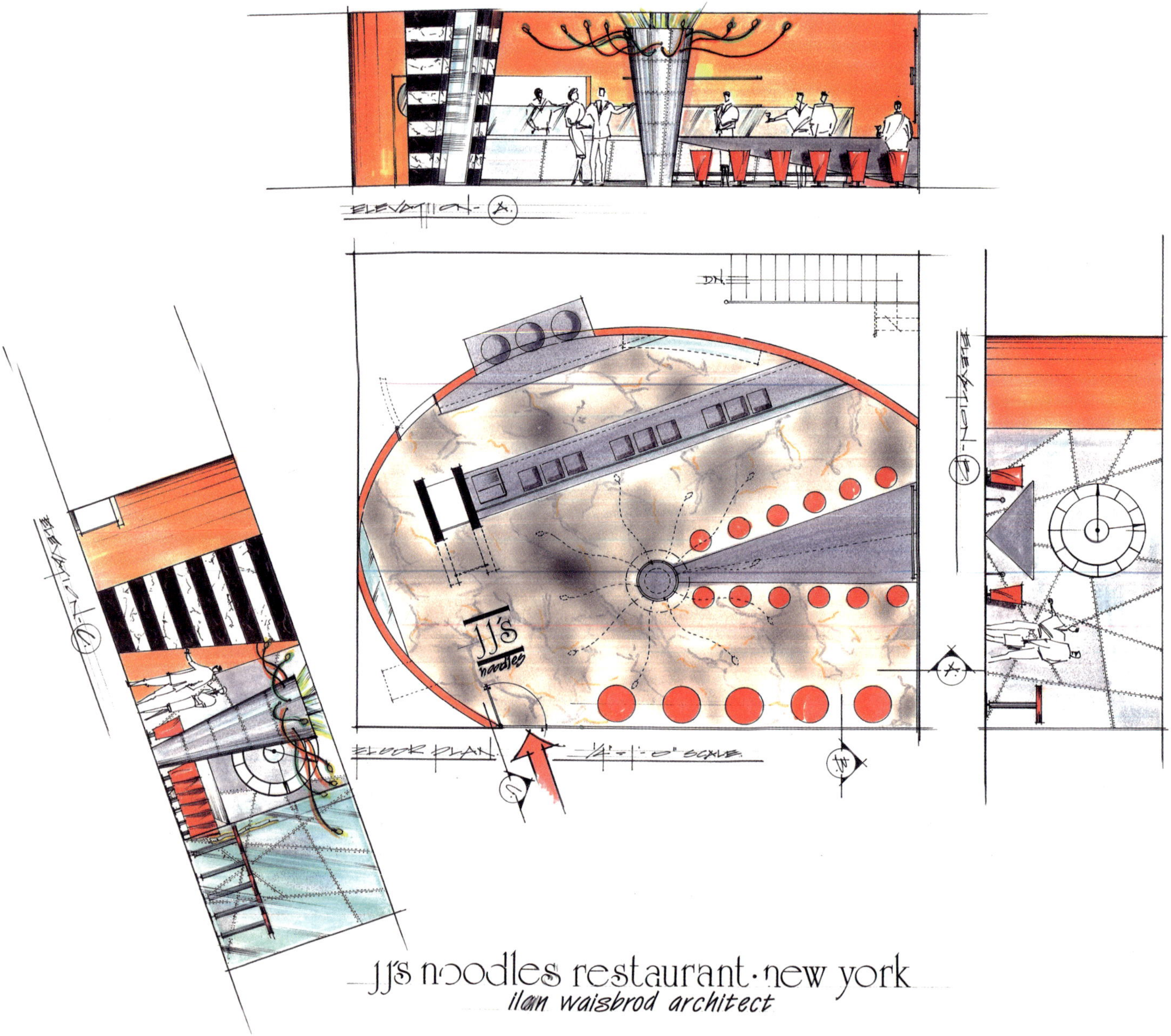

jj's noodles restaurant · new york
ilan waisbrod architect

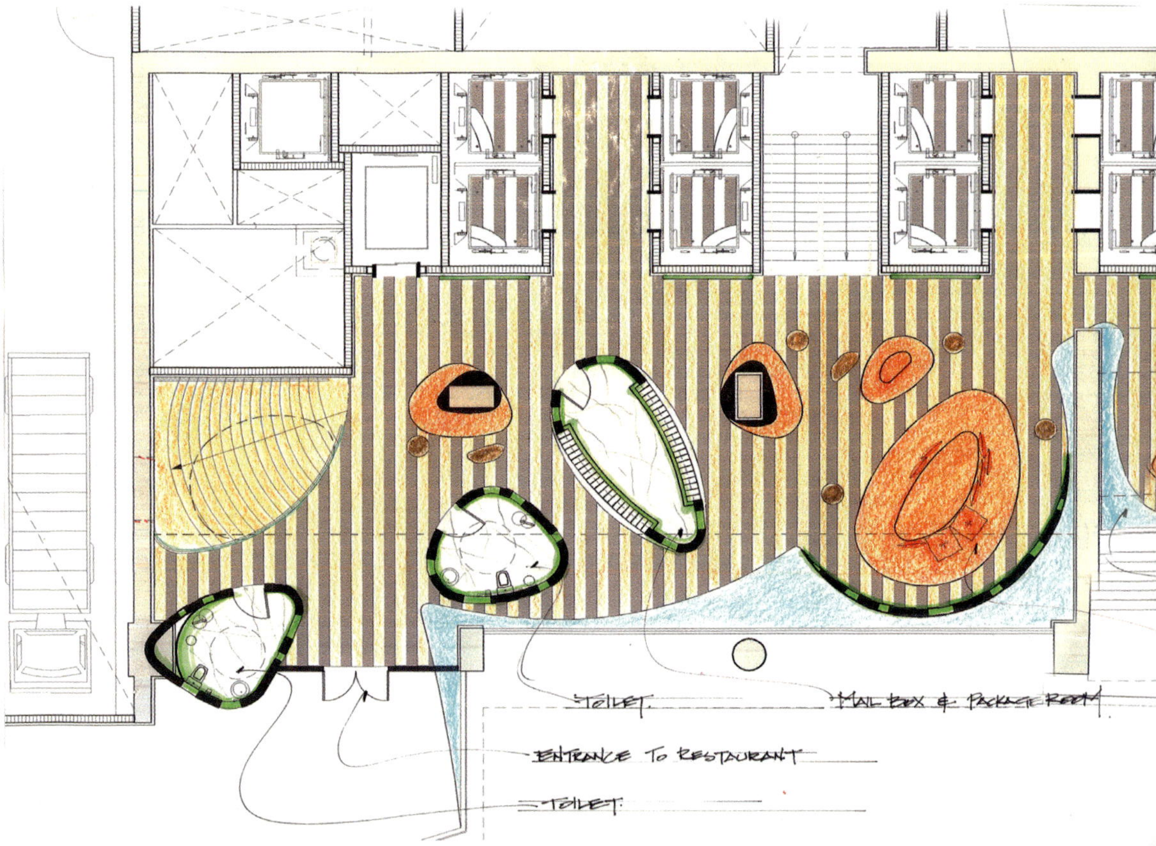
TOILET
MAIL BOX & PACKAGE ROOM
ENTRANCE TO RESTAURANT
TOILET

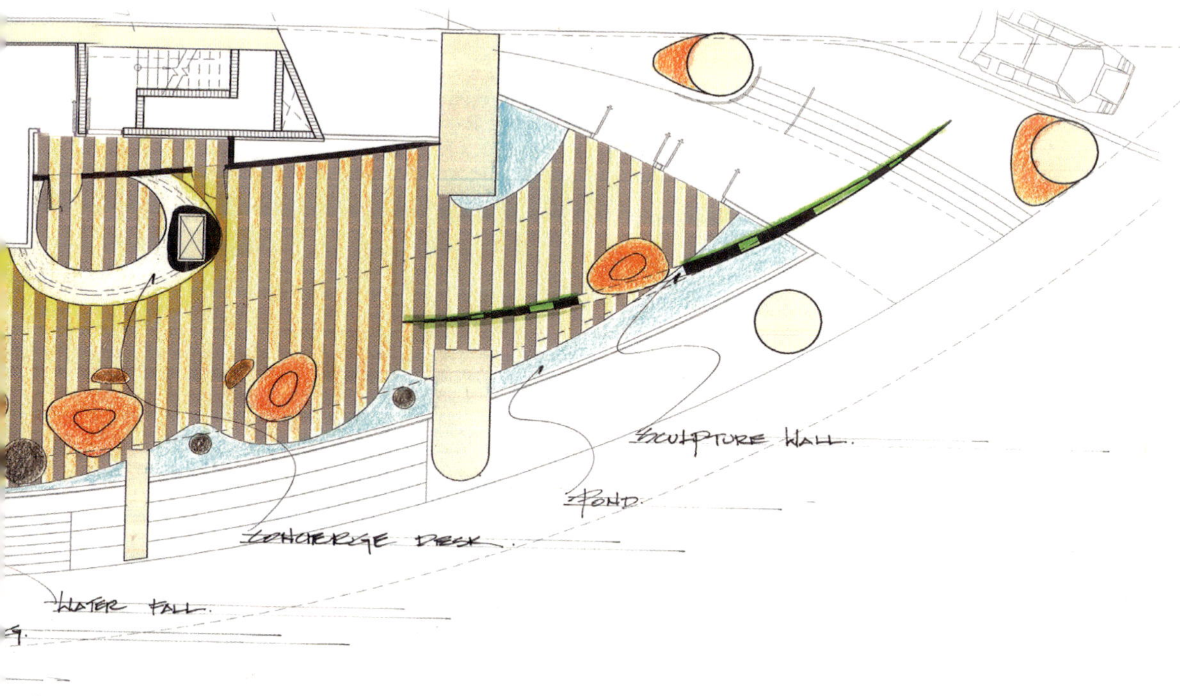
SCULPTURE WALL.
POND.
CONCIERGE DESK.
WATER FALL.

delivery menu

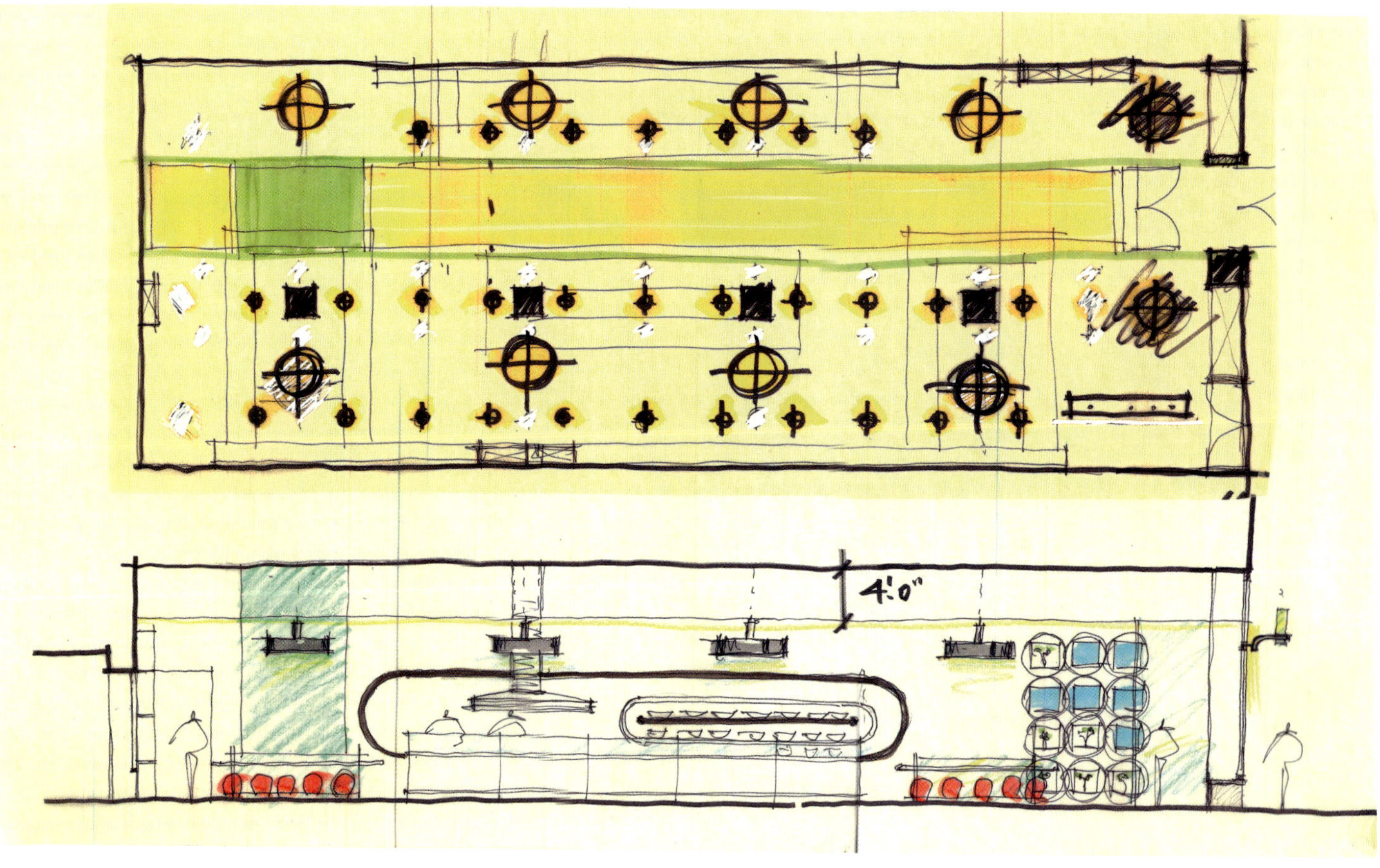
4'-0"

dishes
dishes

dishes
dishes

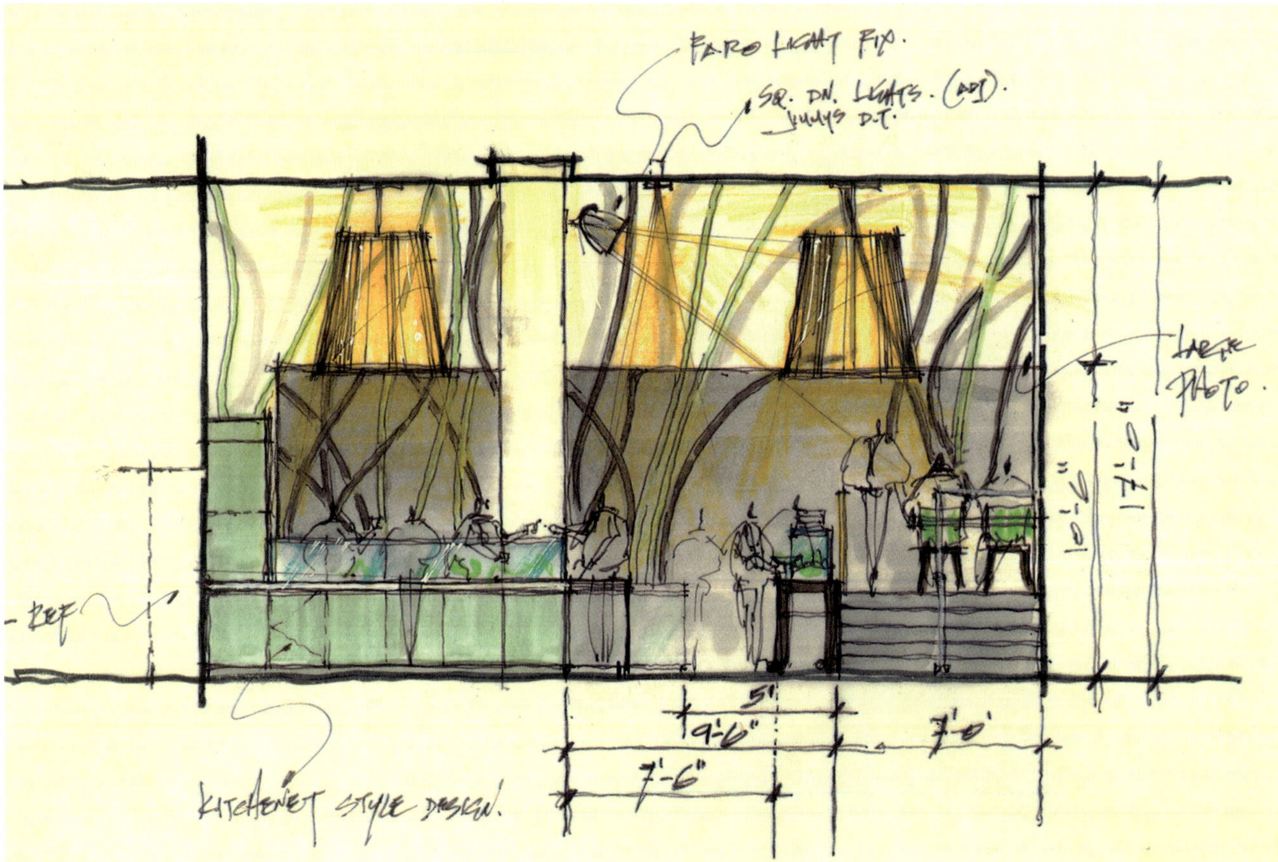

FARO LIGHT FIX.
SQ. DN. LIGHTS.
KITCHENETTE STYLE DESIGN.
LARGE PHOTO.
5'
9'-6"
7'-0"
7'-6"
10'-6"
17'-0"

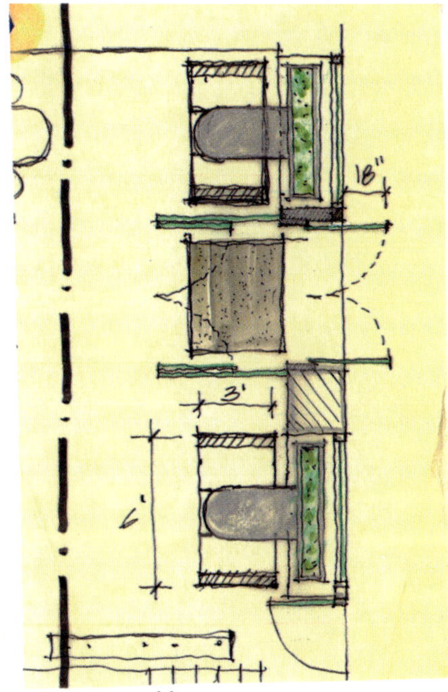
18"
3'
6'

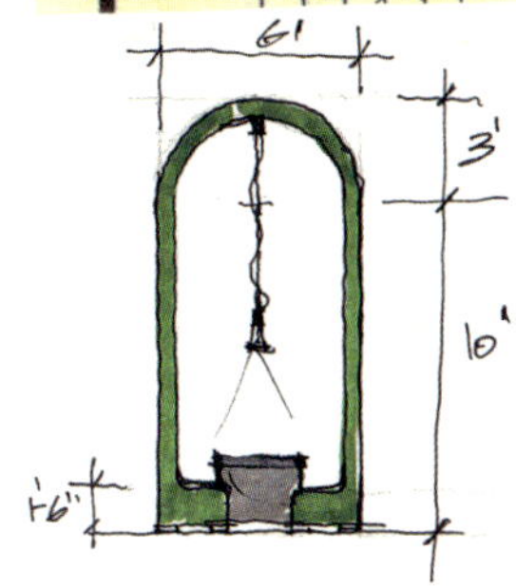
6'
3'
10'
1'6"

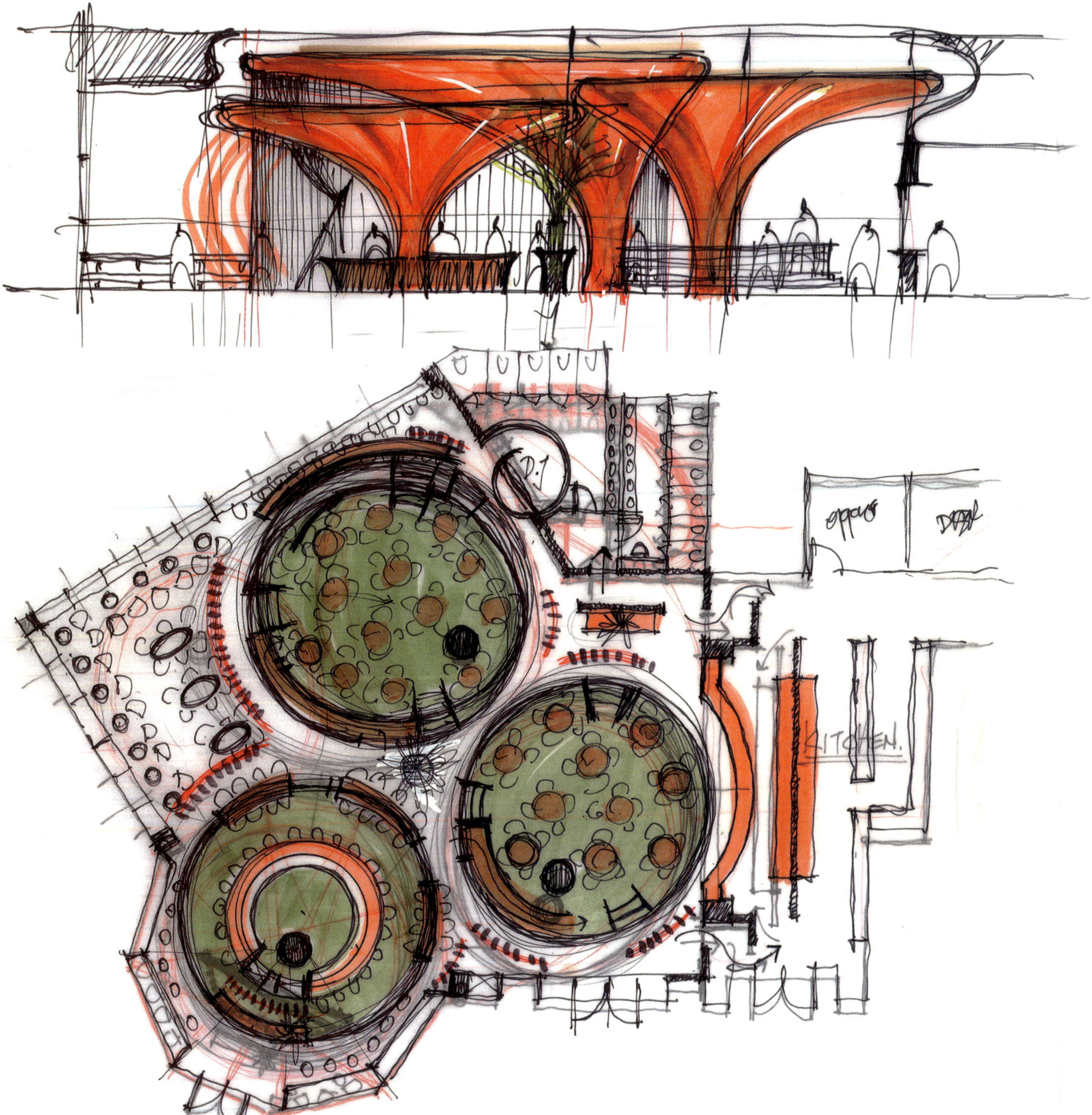
KITCHEN.

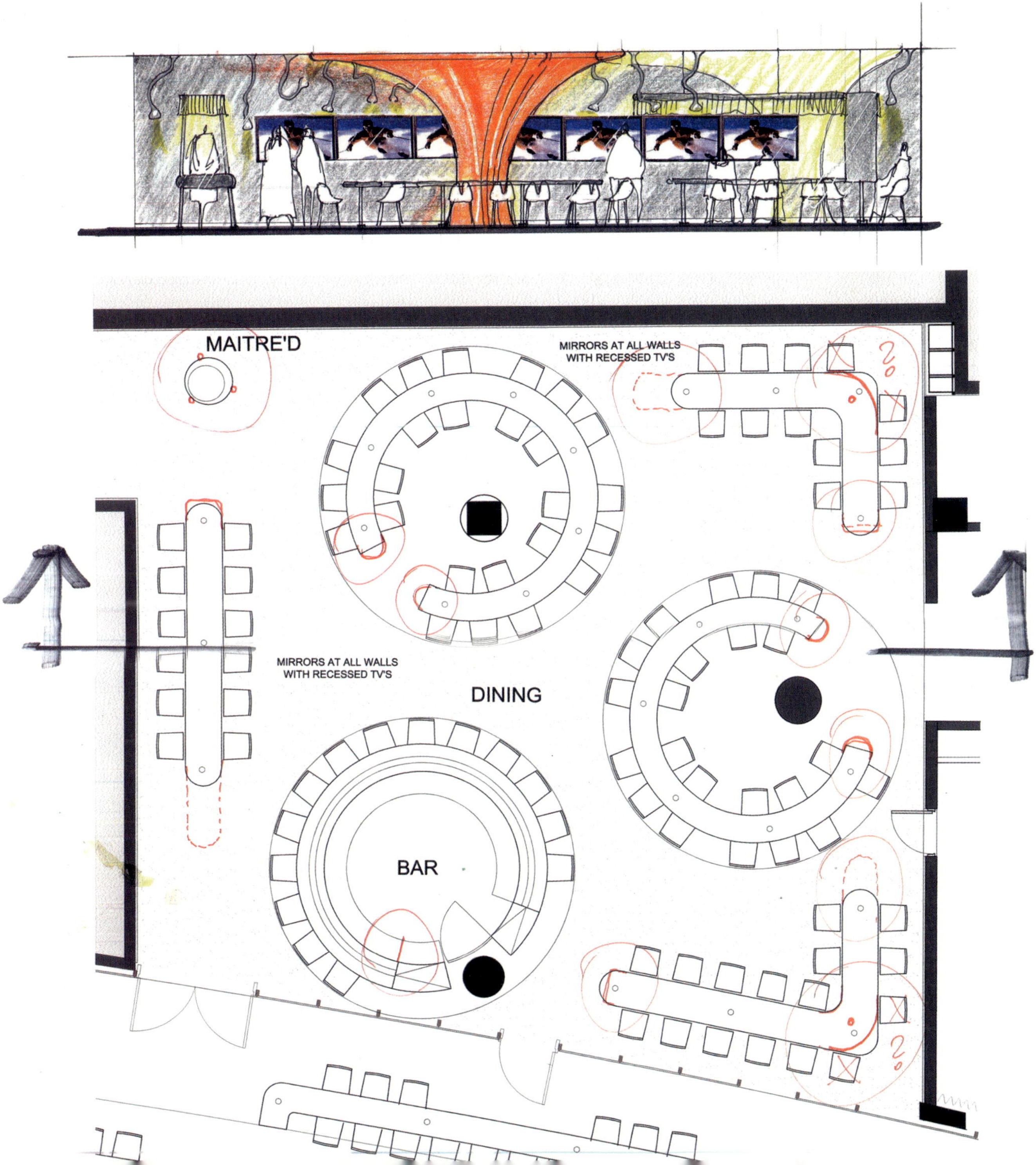
MAITRE'D
MIRRORS AT ALL WALLS
WITH RECESSED TV'S
MIRRORS AT ALL WALLS
WITH RECESSED TV'S
DINING
BAR

ILAN 2009

CUSTOMER WINE/WHISKEY STORAGE
WOMENS BATHRM
PAINTING
VEST
STAGE
PLASMA TV ON WALL
LOBBY ENTRANCE
PLASMA TV ON WALL
CASHIER/MAITRED STAND
CURTAINS, TYP.
SUNKEN LOUNGE SEATING
DISPLAY CABINET
STOR
CASINO RECEPTION
HOTEL RECEPTION
UP
DN
DELI
CORRIDOR
STOR
STOR

ELEVATION

INTERIOR ELEVATION
SCALE=1:50

SCULPTURE
BAR
VIDEO WALL
SCULPTURES
CONSERGE DESK
STORAG

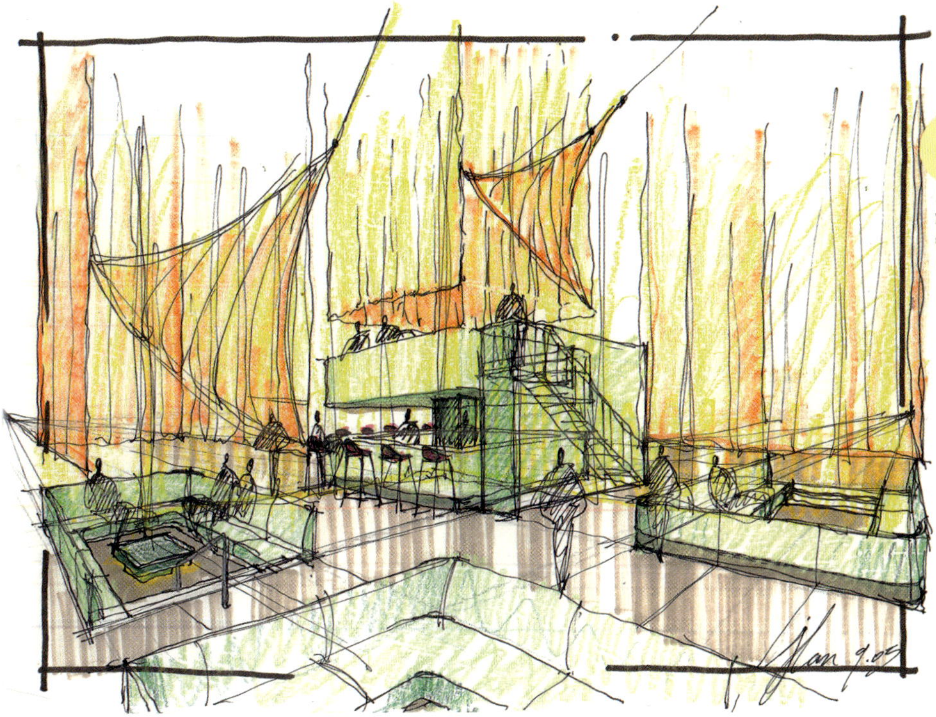

4'
12'
DN.
DN.
8'
2'
2'
12'

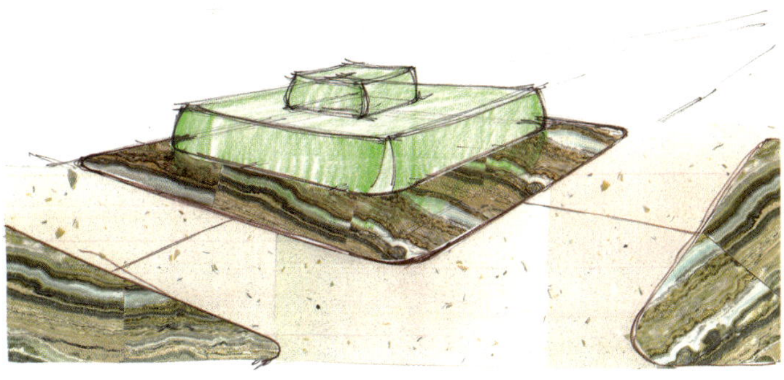

ELEVATOR CORRIDOR

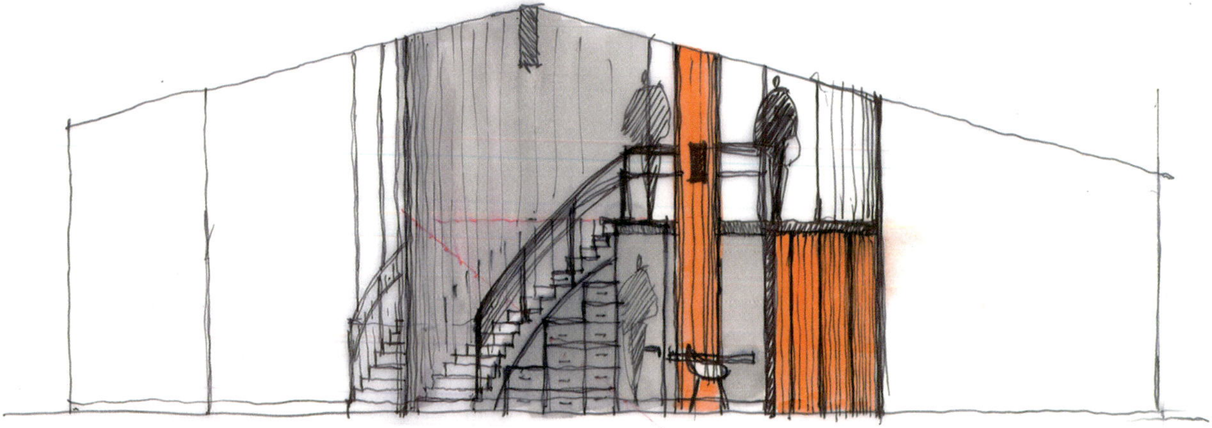

EL.= 730'-0"
9'-0"
1'-9"
3'-0"

THE RELATED COMPANIES LP
RANDOM HOUSE INC.
Skidmore Owings & Merrill LLP
Ismael Leyva Architects P.C.
Thornton Tomasetti
Cosentini Associates
Vollmer Associates
Langan Engineering PC
Shen Milsom Wilke
Philip Habib and Associates
EL.= 733'-0"
EL.= 719'-0"
16" dia flue
8'-3"
50th Mezz. 1
EL.= 712'-8"
50th Floor
EL.= 701'-8"

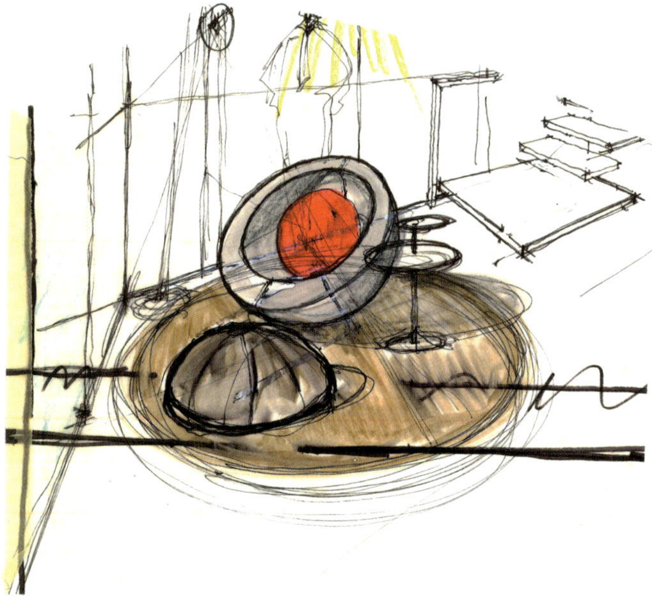

Coffee

Coffe
reto
Desert
Juice
Ice cream
Bakery
Panini
Coffe
Cash
trash

SALADS

reto
Bakery
coffe
deserts
Salads
reto
ice cream
Bokery
cash
reto

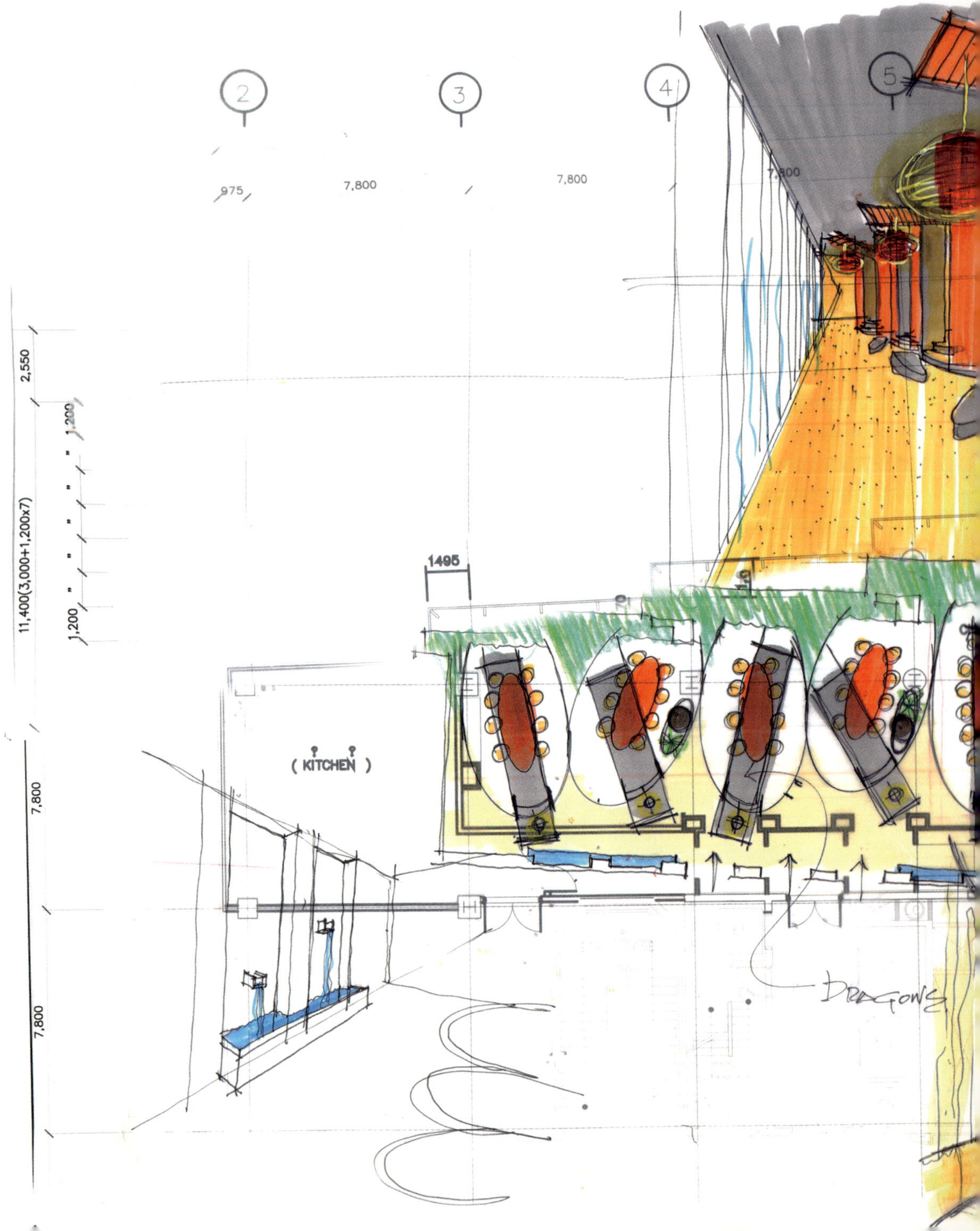

2
3
4
5
975
7,800
7,800
7,800
2,550
1,200
11,400(3,000+1,200x7)
1,200
1495
(KITCHEN)
7,800
7,800
DRAGONS

7
8
9
7,800
7,800
7,800

MAIN SEATING AREA. (±65 SEATS).
M'D STAND.
DRINK RAIL.
DANCE FLOOR.
TOILETS
PROJECTION ON WATER FALL.
D.J.
BAR.
PICK UP AREA.
STANDING BAR.
DANCE FLOOR.
PICK UP AREA.
OUT DOOR SEATING (±55)
PRIVATE SEATING. (±15
PRIVATE SEATING (±15)
PRIVATE SEATING. (±15).
DN
4056
4004
4003
6
7

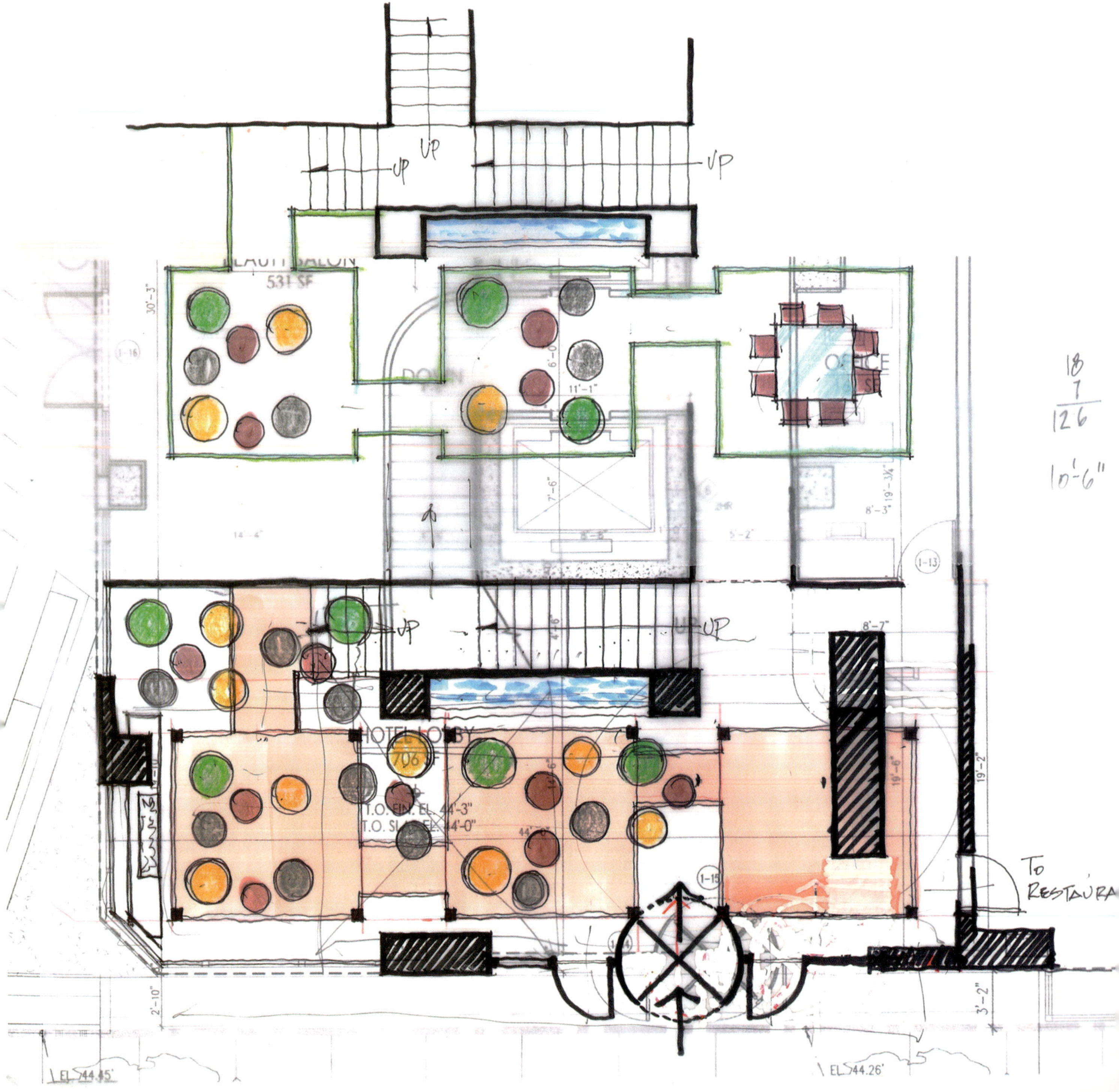

UP
UP
UP
531 SF
30'-3"
DOWN
6'-0"
11'-1"
OFFICE
18
7
126
10'-6"
7'-6"
14'-4"
8'-3"
5'-2"
UP
8'-7"
HOTEL LOBBY
706 SF
T.O. FIN. EL. 44'-3"
19'-6"
19'-2"
To
RESTAURA
2'-10"
3'-2"
EL 44.45'
EL 44.26'

ILAN. 2008.

TATI

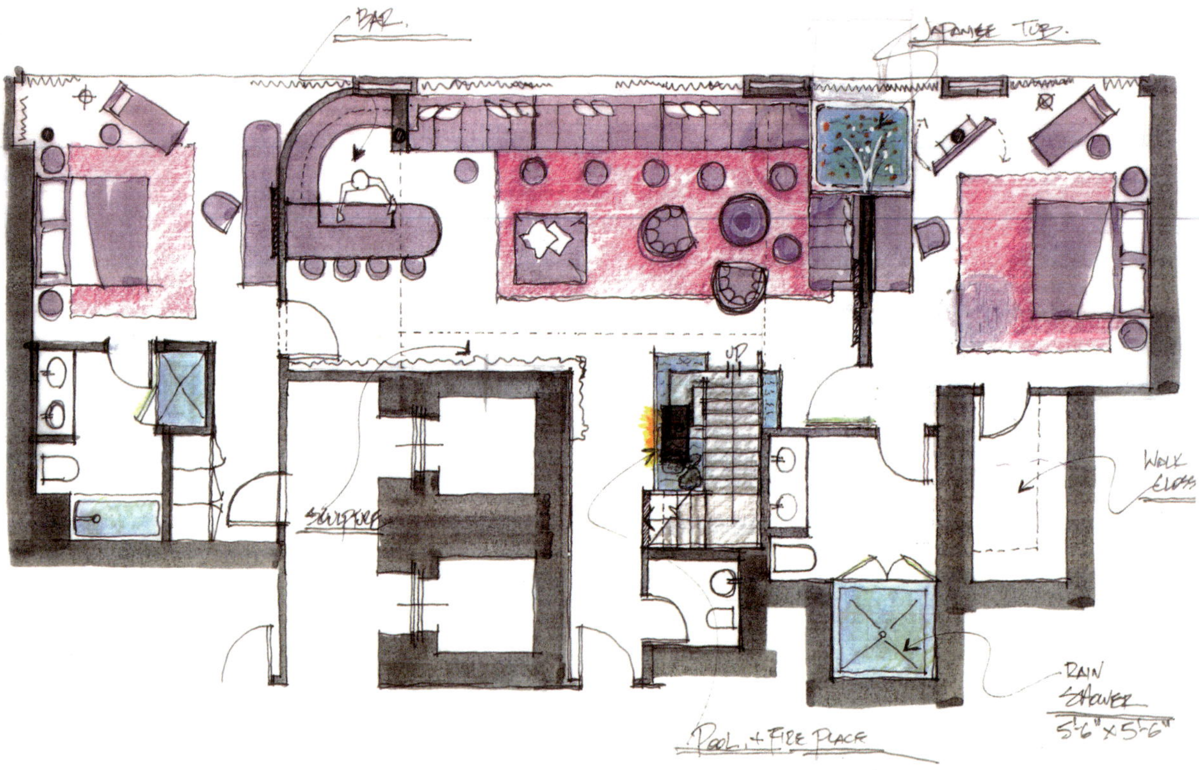
BAR.
JAPANESE TUB.
UP
WALK CLOSS
SCULPTURE
RAIN SHOWER
5'-6" x 5'-6"
POOL + FIRE PLACE

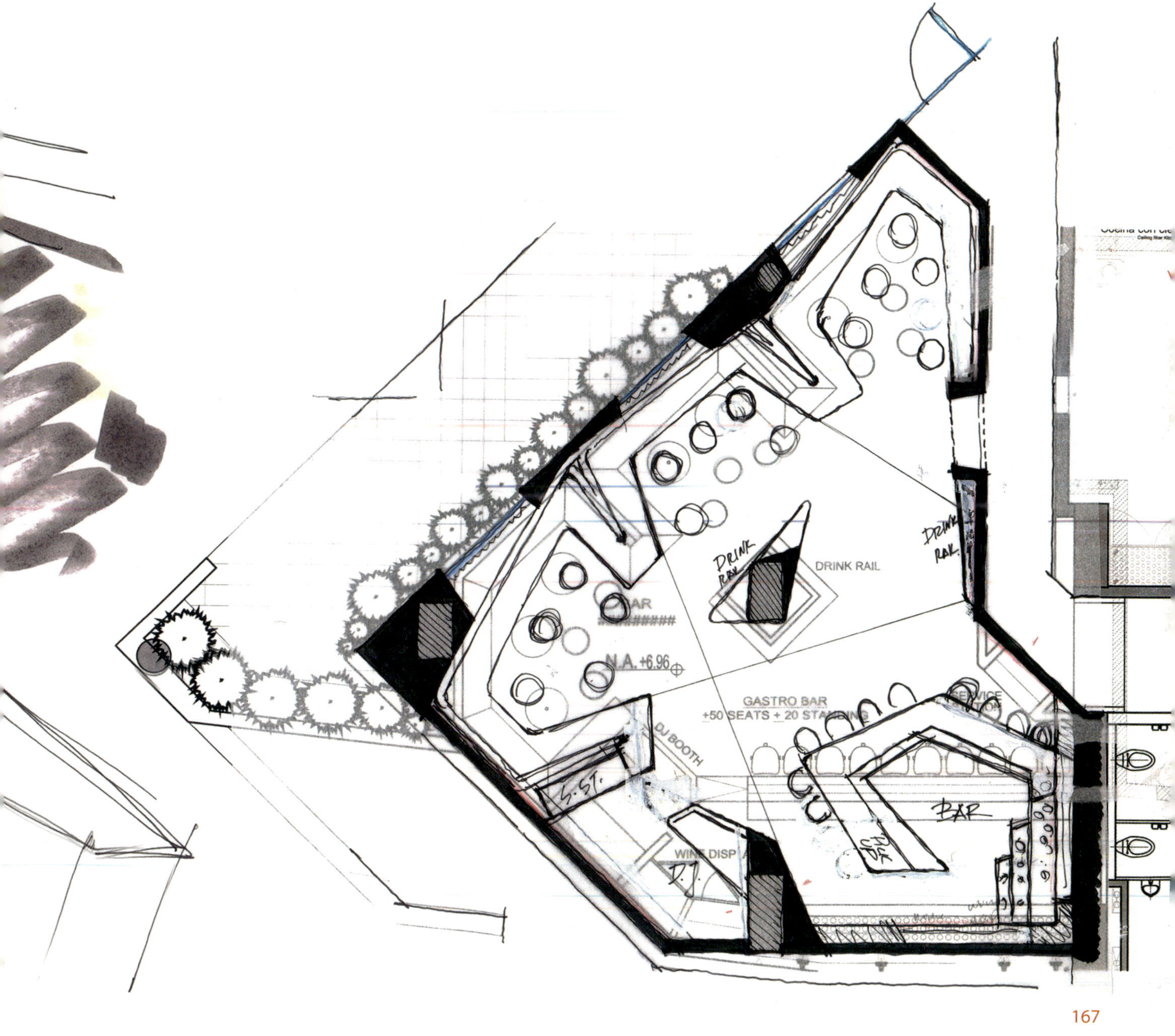
DRINK RAIL
DRINK RAIL
DRINK RAIL
N.A. +6.96
GASTRO BAR
+50 SEATS + 20 STANDING
DJ BOOTH
BAR
WINE DISP

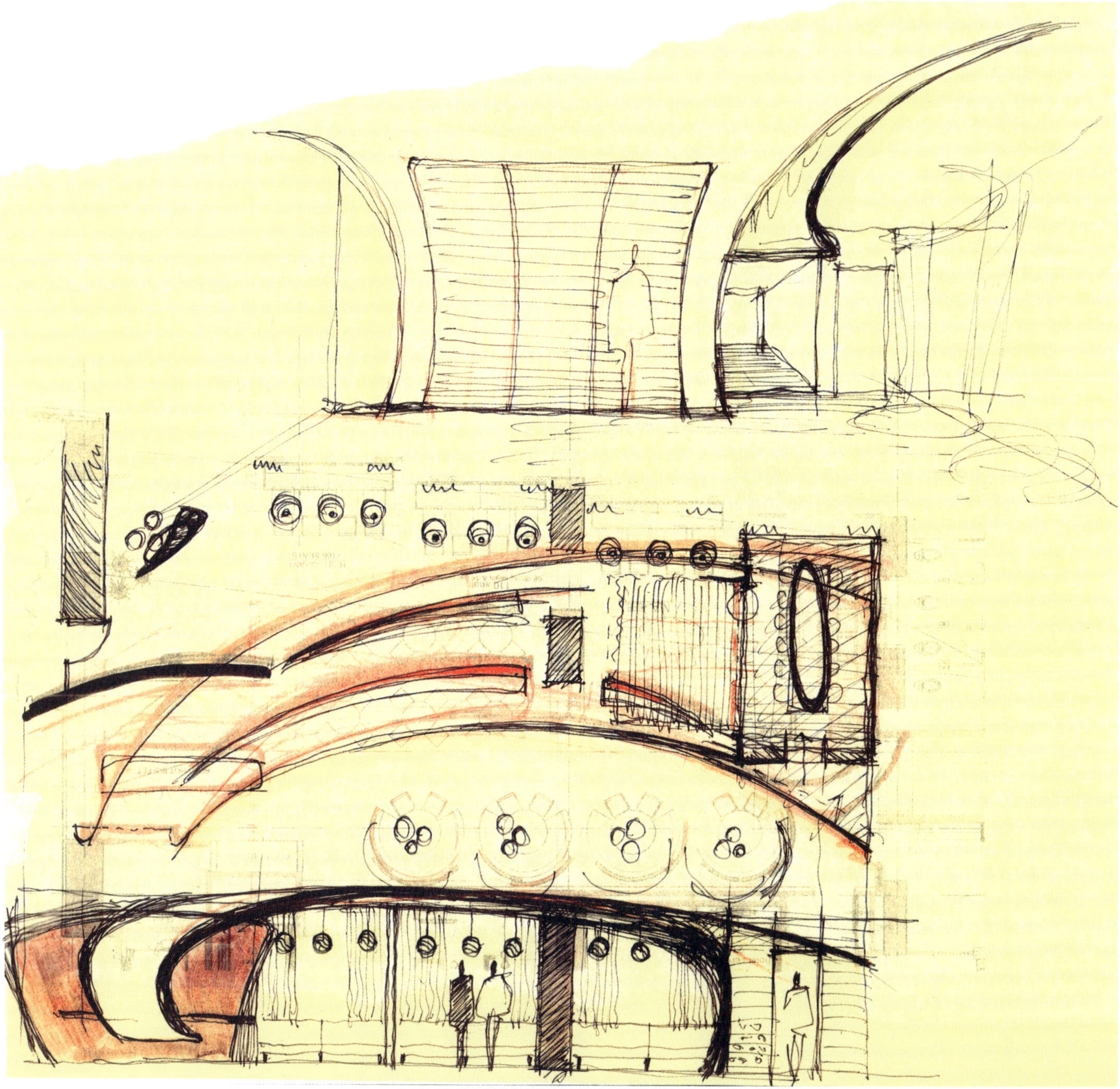

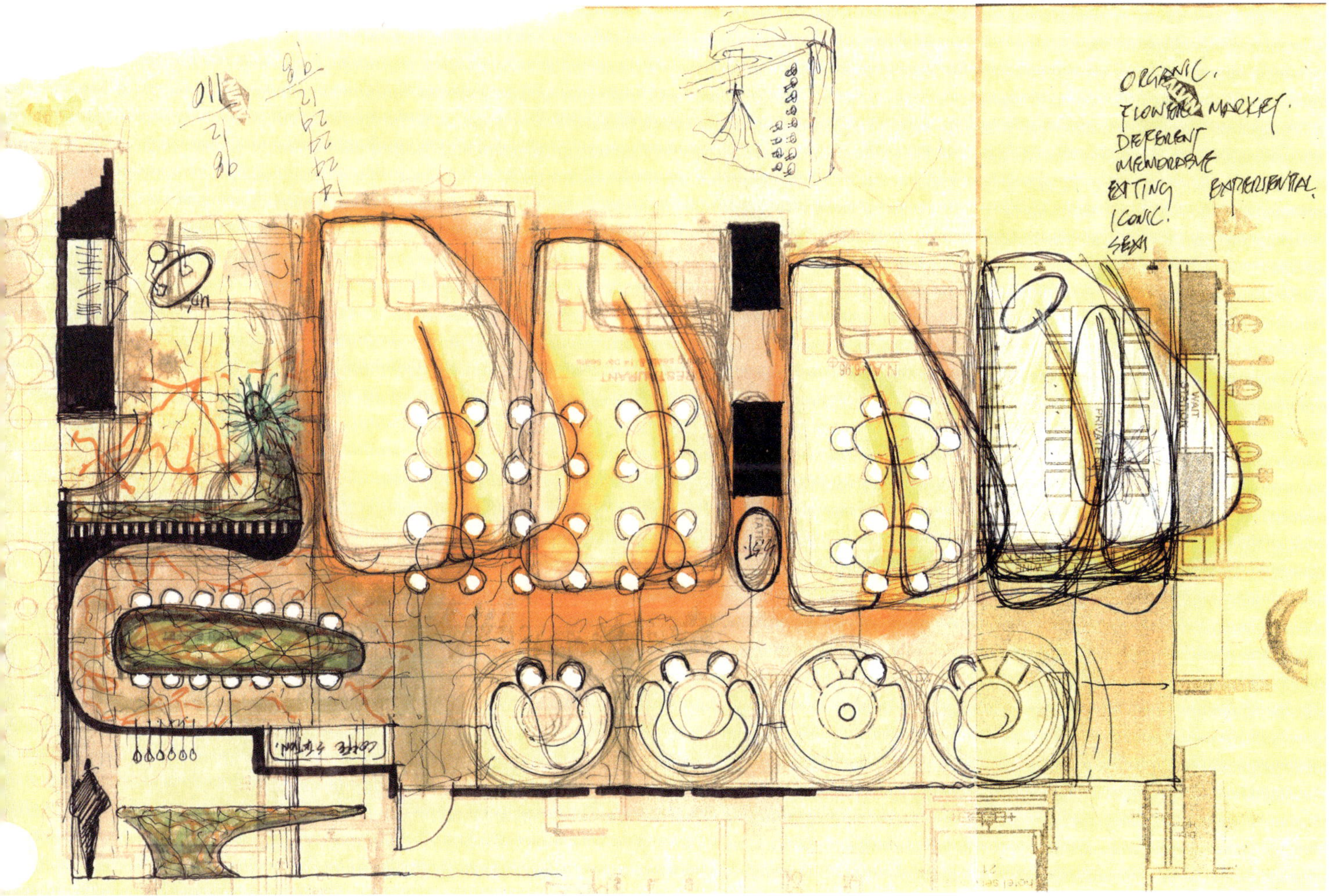
ORGANIC.
FLOWER MARKET.
DIFFERENT
MEMORABLE
EXITING
EXPERIENTIAL.
ICONIC.
SEXY
PRIVATE

RAINFOREST GREEN
MARBLE.

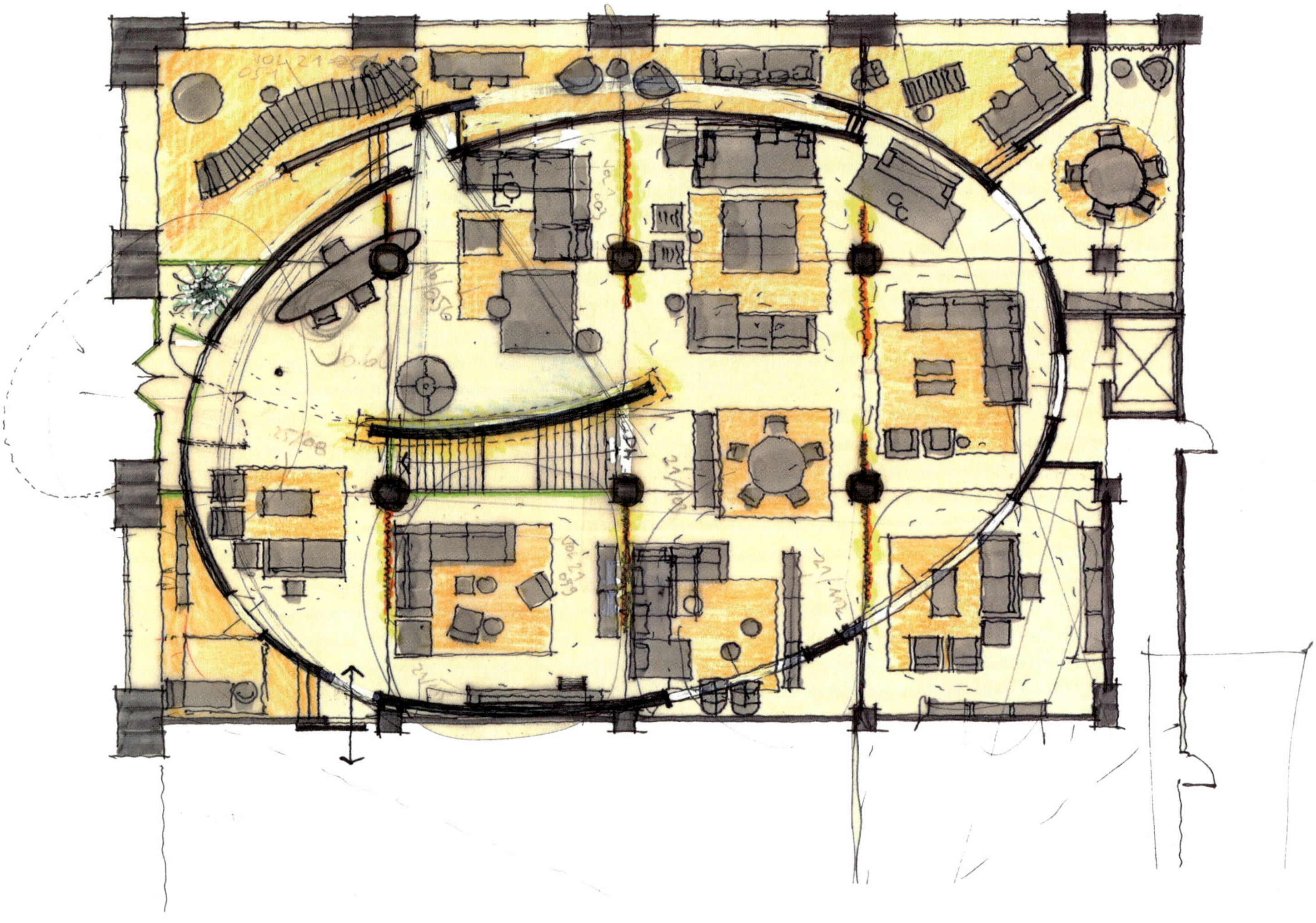

DDC SHOWROOM, N.Y. 20

KITCHEN

YUNG & CHEERFULL
WARM & LIVELY.

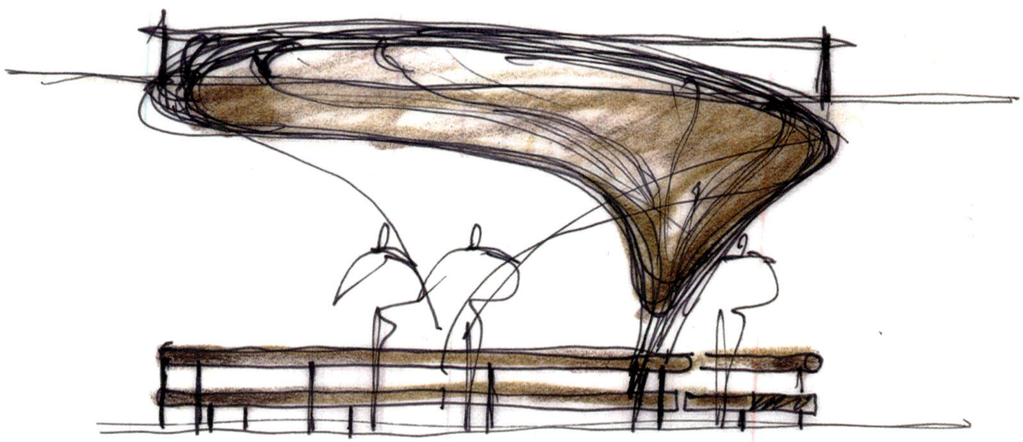

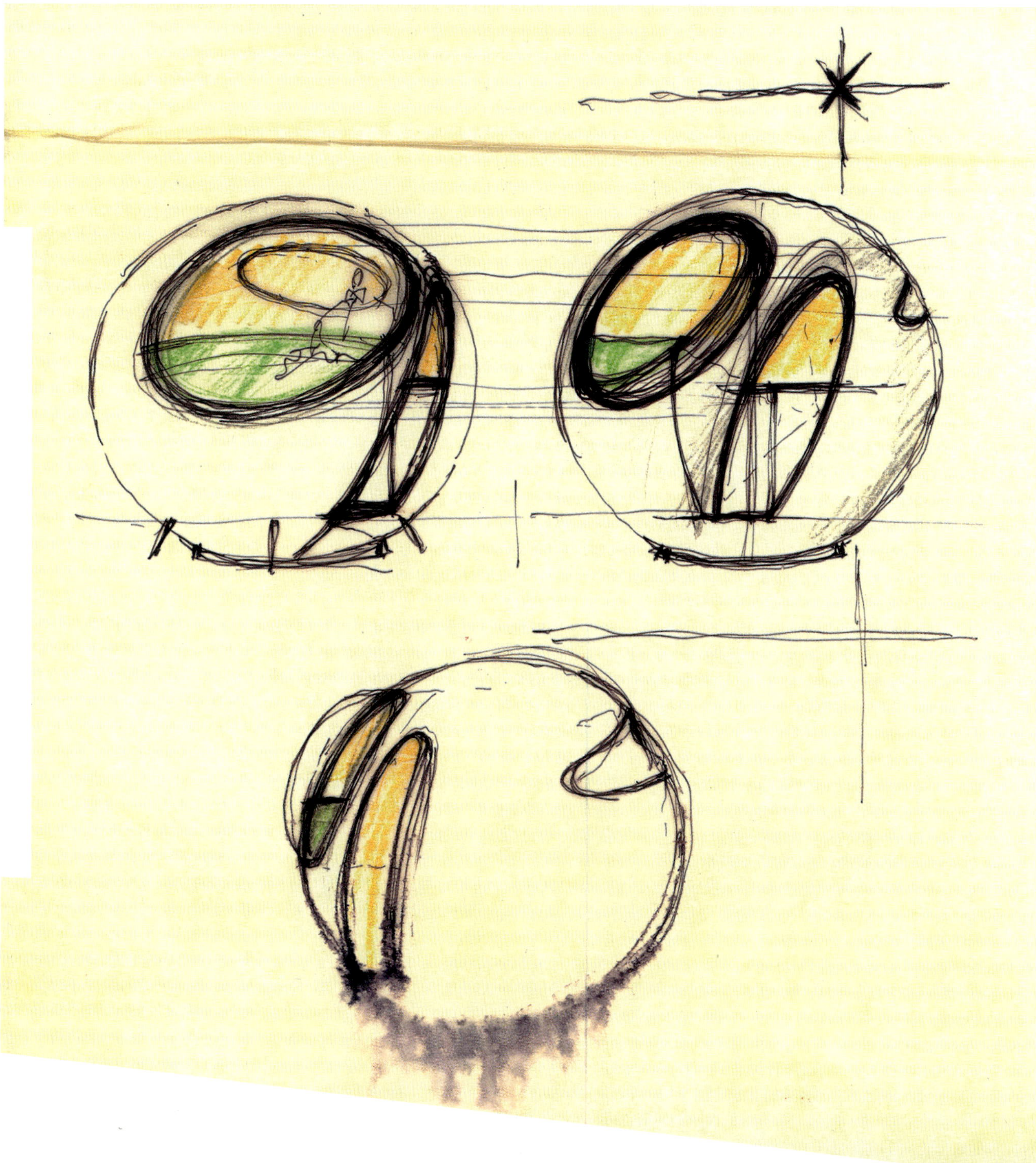

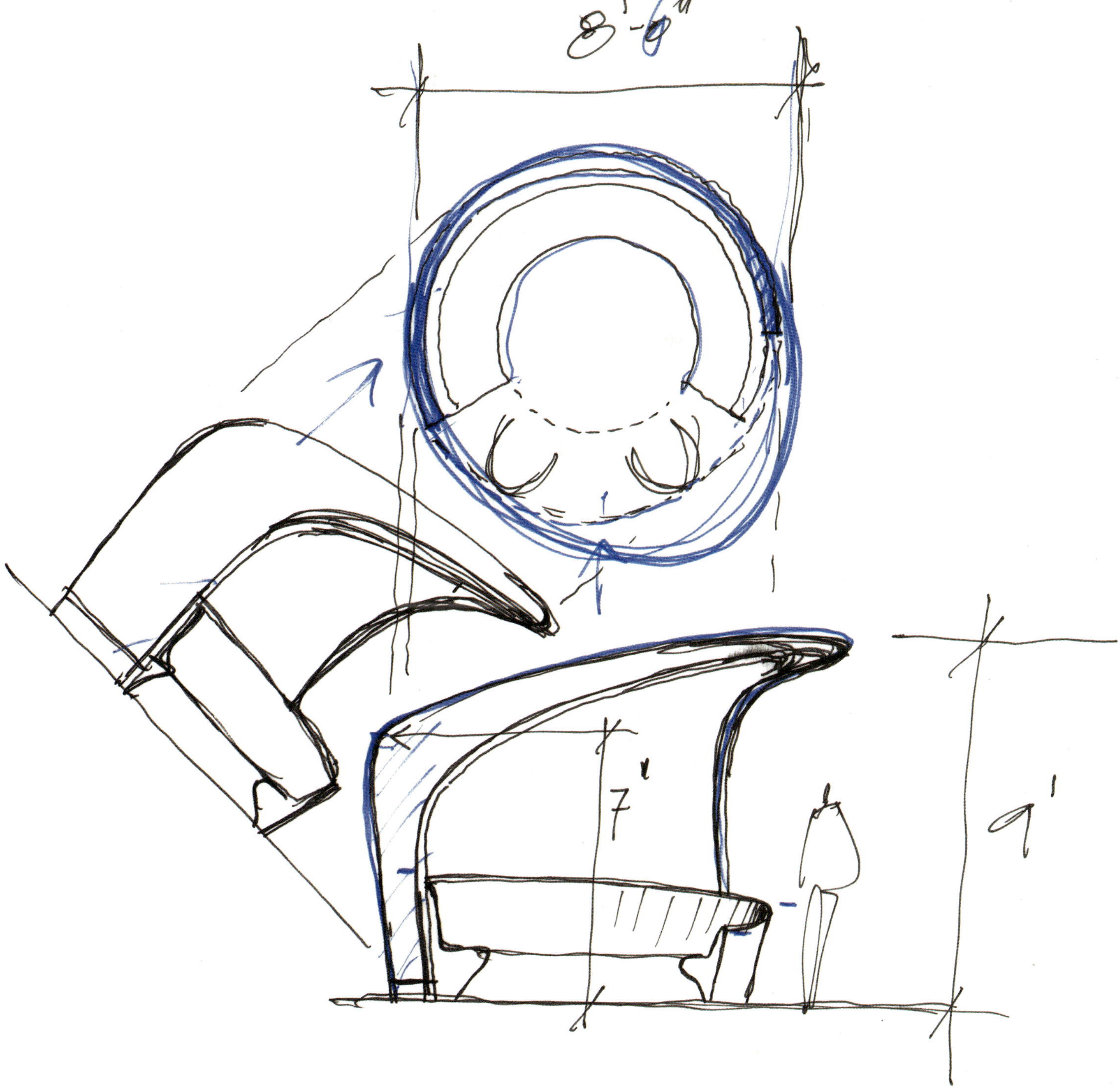
8'-6"
7'
9'

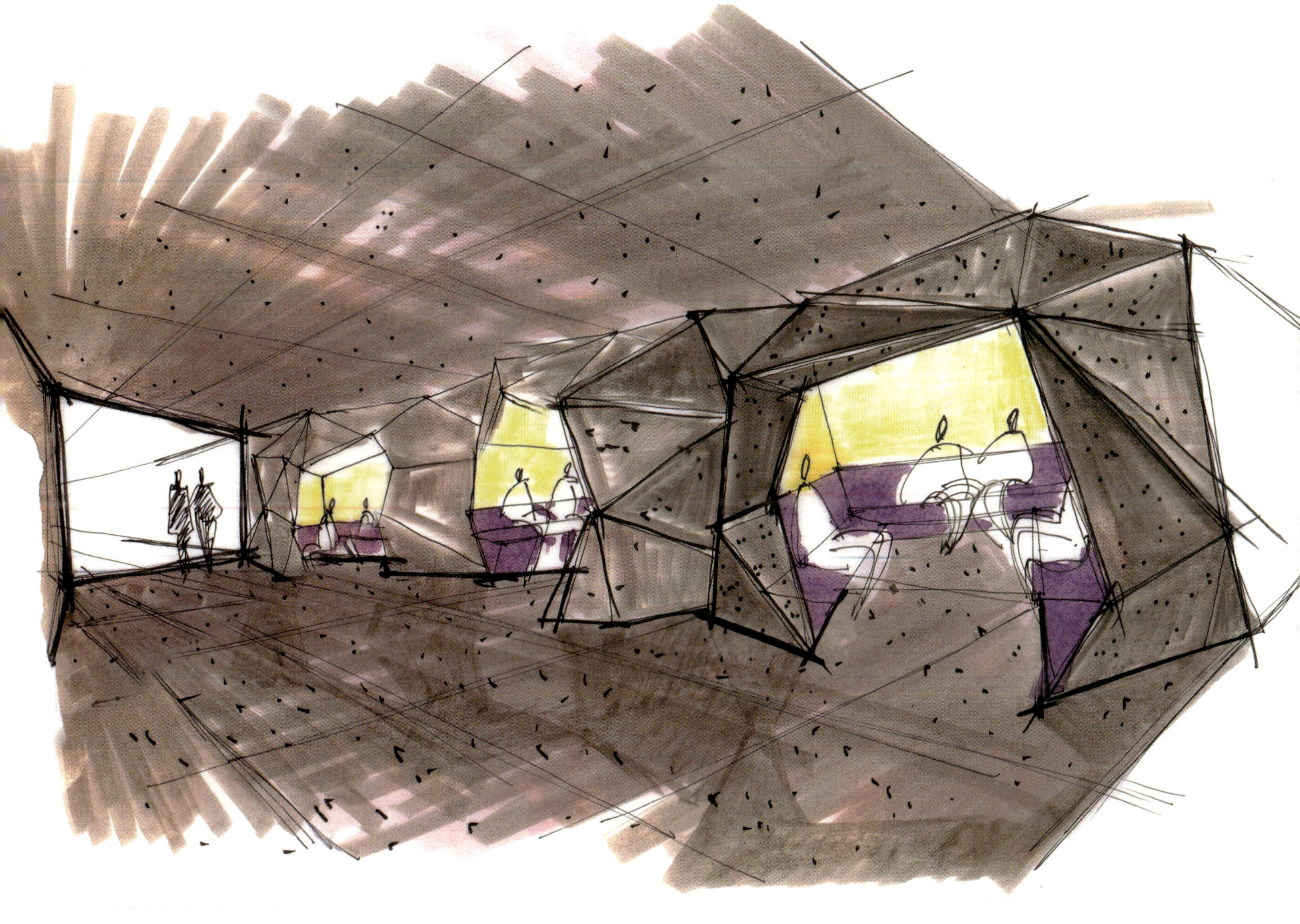

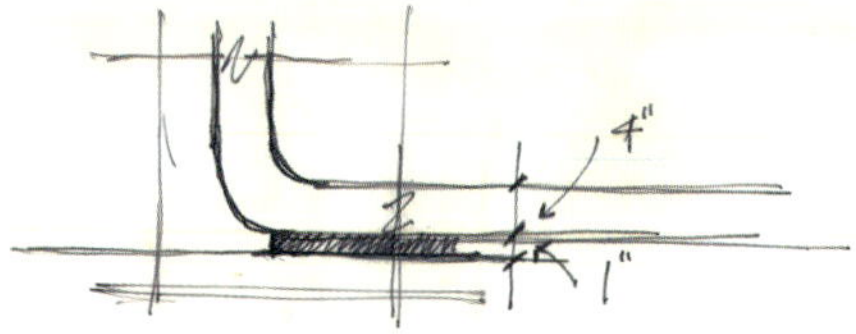
4"
1"

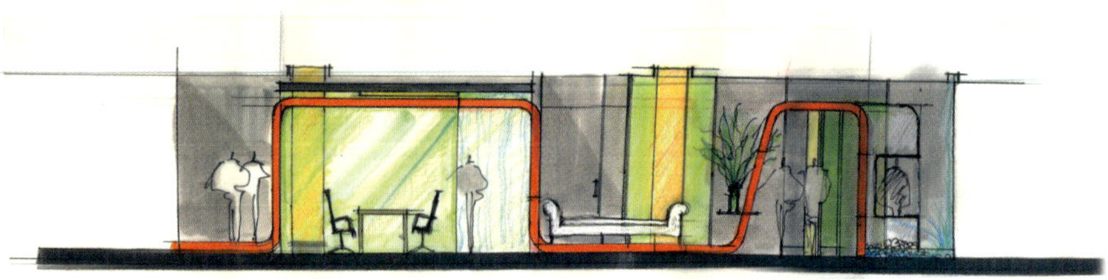

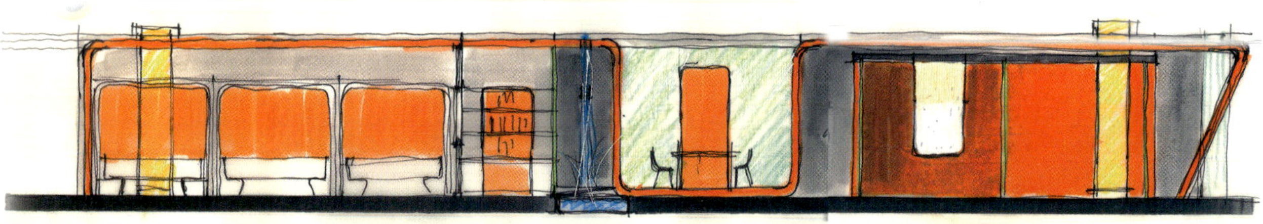

ENTRANCE TO
MAIL ROOM.

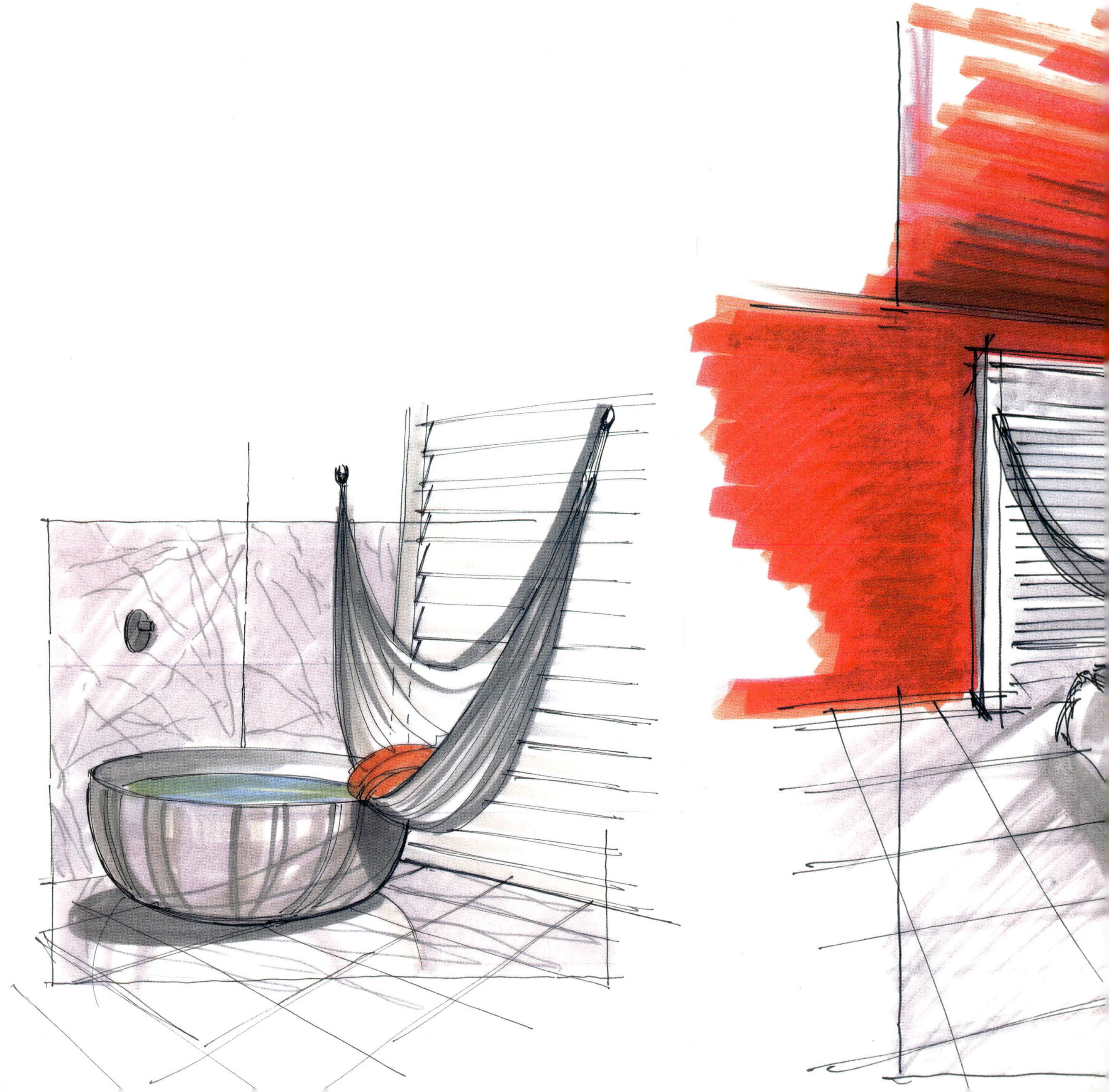

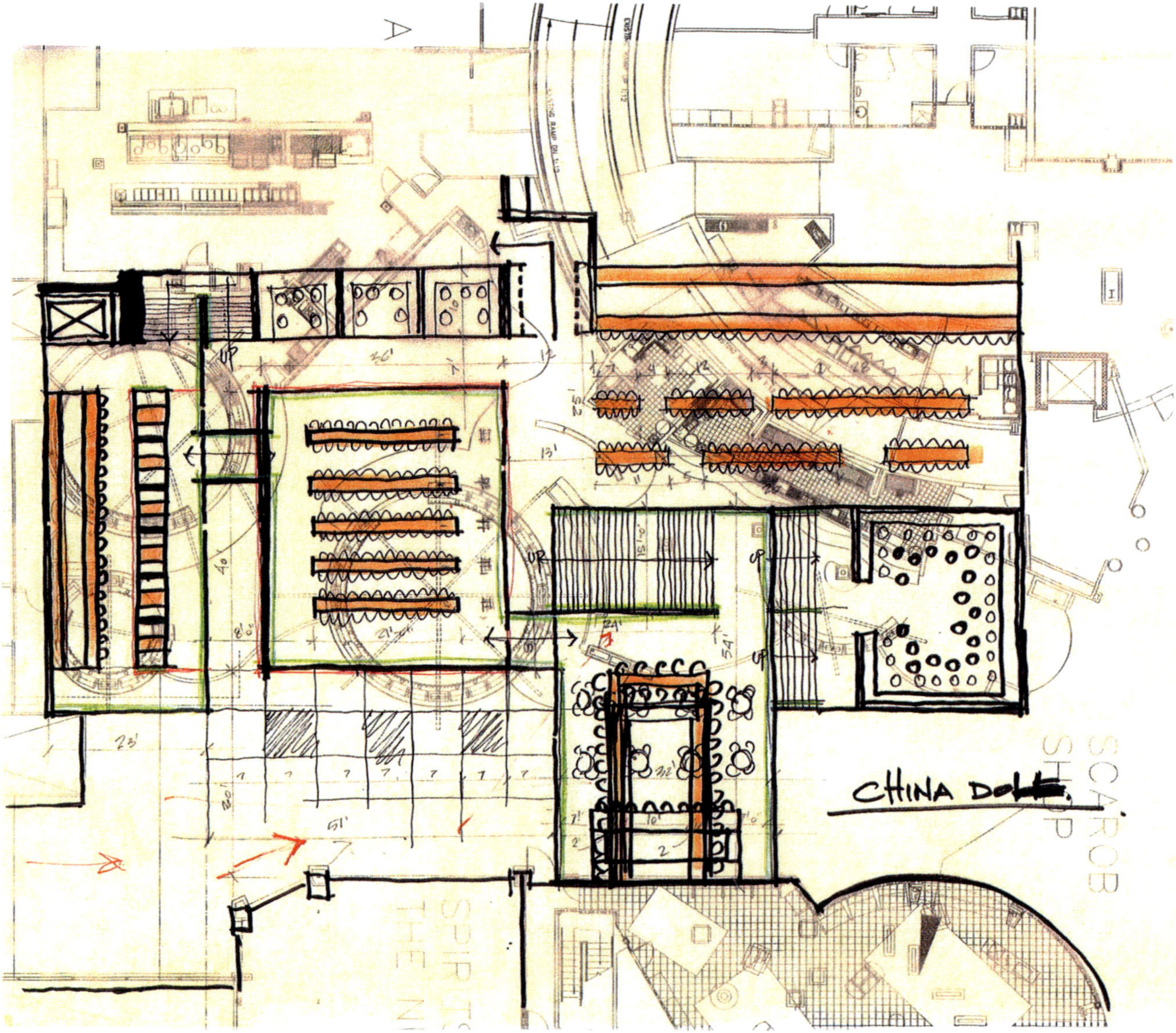
CHINA DOLL

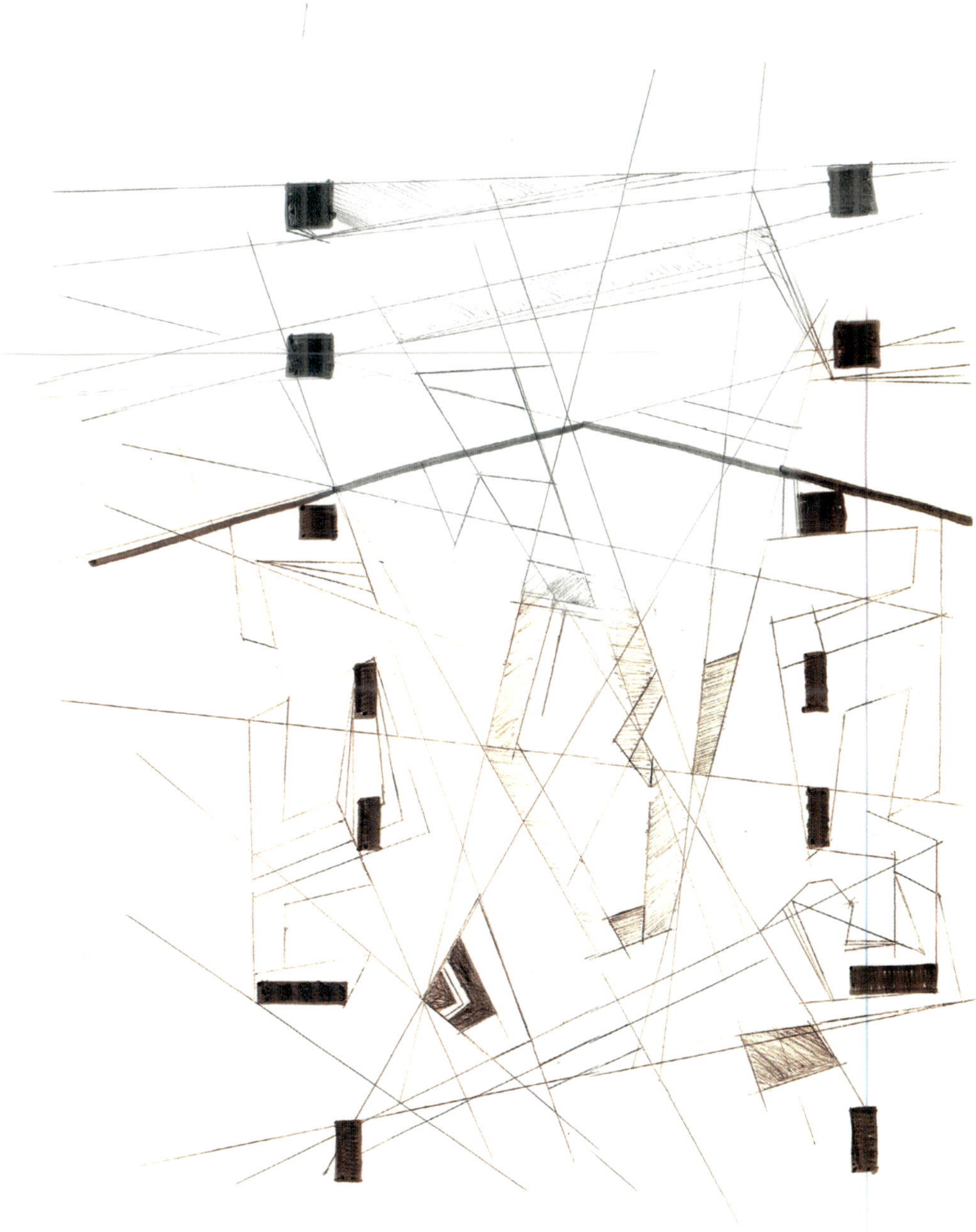

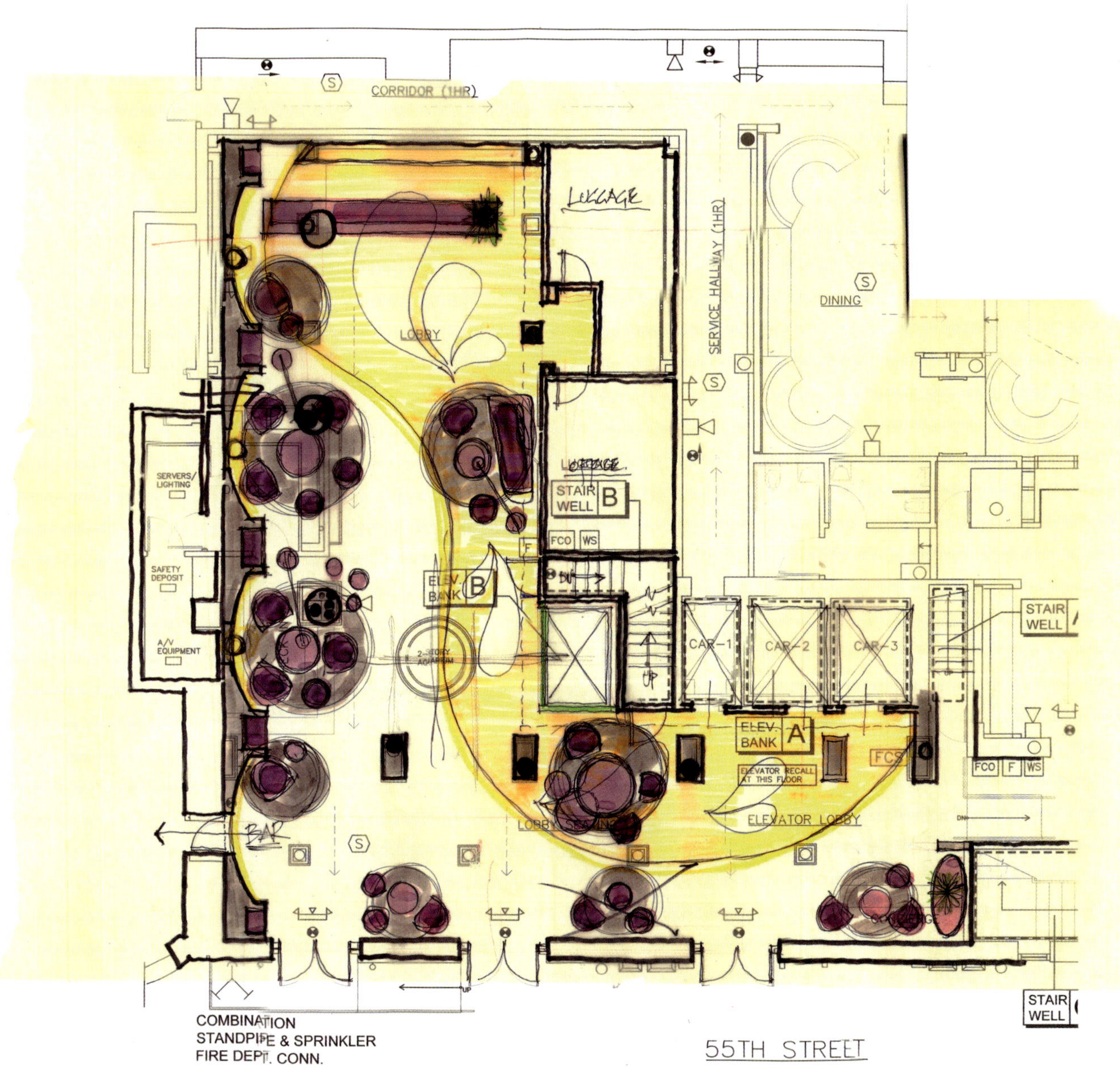
CORRIDOR (1HR)
LUGGAGE
SERVICE HALLWAY (1HR)
DINING
LOBBY
SERVERS/ LIGHTING
SAFETY DEPOSIT
A/V EQUIPMENT
STAIR WELL B
FCO
WS
ELEV. BANK B
CAR-1
CAR-2
CAR-3
STAIR WELL
ELEV. BANK A
ELEVATOR RECALL AT THIS FLOOR
ELEVATOR LOBBY
LOBBY SEATING
BAR
STAIR WELL
COMBINATION
STANDPIPE & SPRINKLER
FIRE DEPT. CONN.
55TH STREET

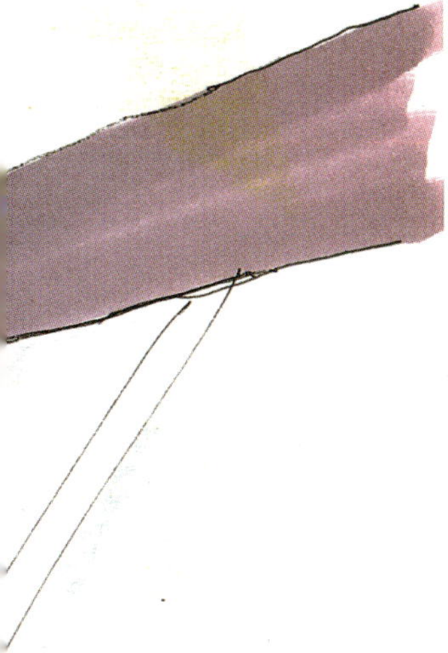

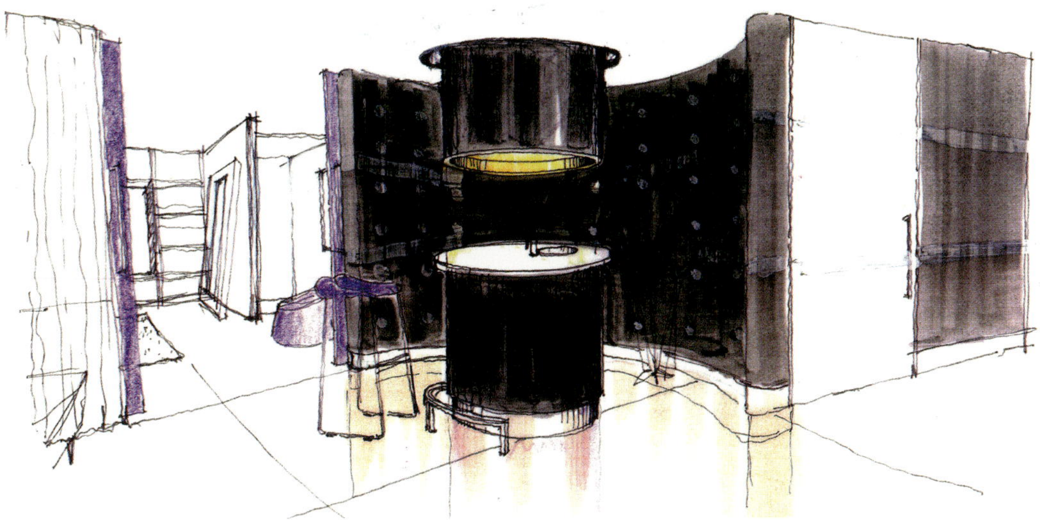

TOILET.
POWDER ROOM.
DUBLE SINK.
STONE, FEATURE WALL.
MINIBAR & SHELVES.
FREE STANDING TUB
SHOWER.
BAR COUNTER + STOOLS
LONG MIRROR, T.V.
TERAZO FLOOR.
GOLF PAD.
5'Ø CHANDELIER
BUILT IN CABINET.

B.

GUCCI

delightful

ORANGE TILE.

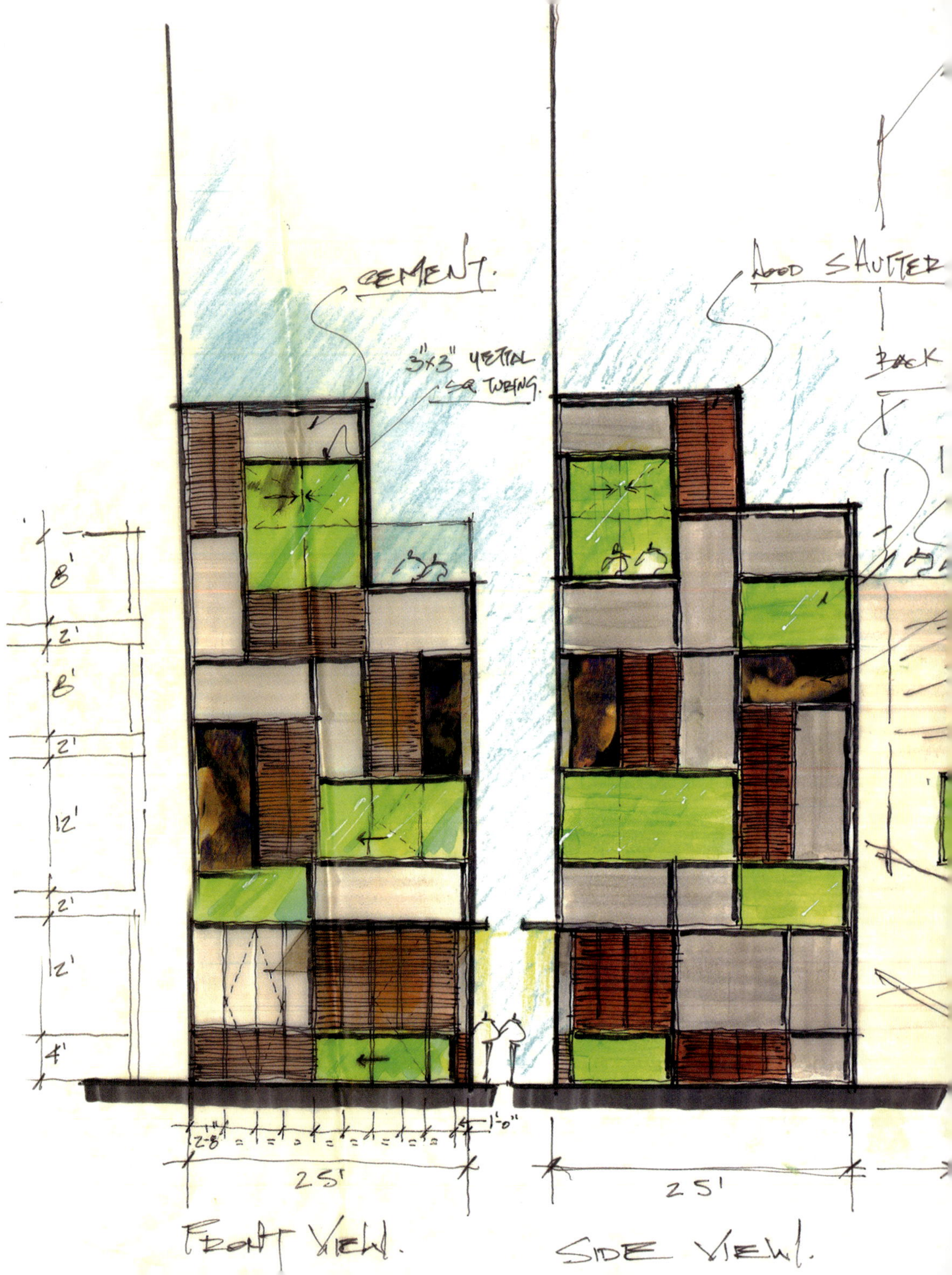
CEMENT.
3"x3" METAL
SQ TUBING.
WOOD SHUTTER
BACK
8'
2'
8'
2'
12'
2'
12'
4'
2'-8"
1'-0"
25'
25'
FRONT VIEW.
SIDE VIEW.

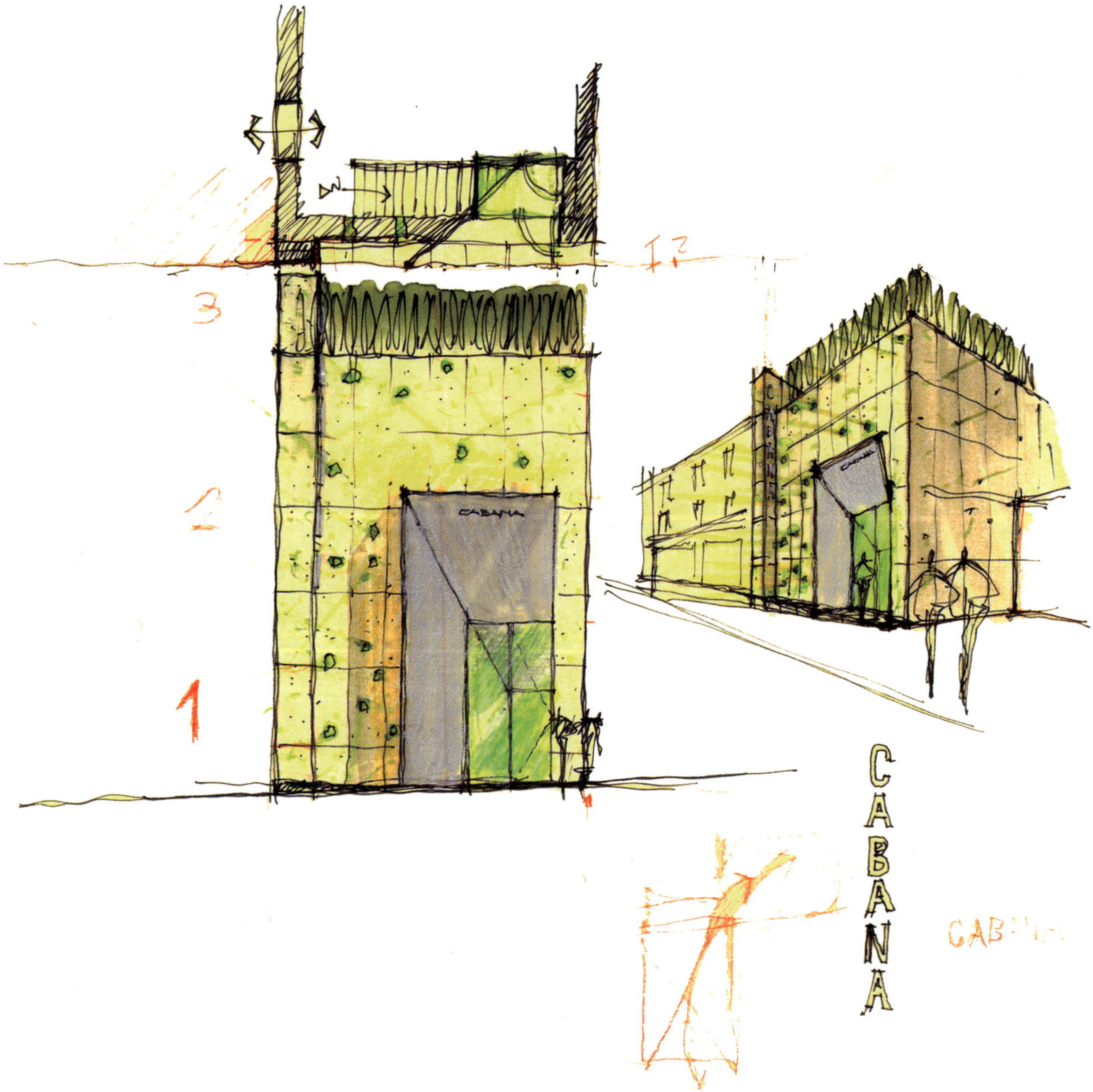
DN
3
2
1
CABANA
CABANA

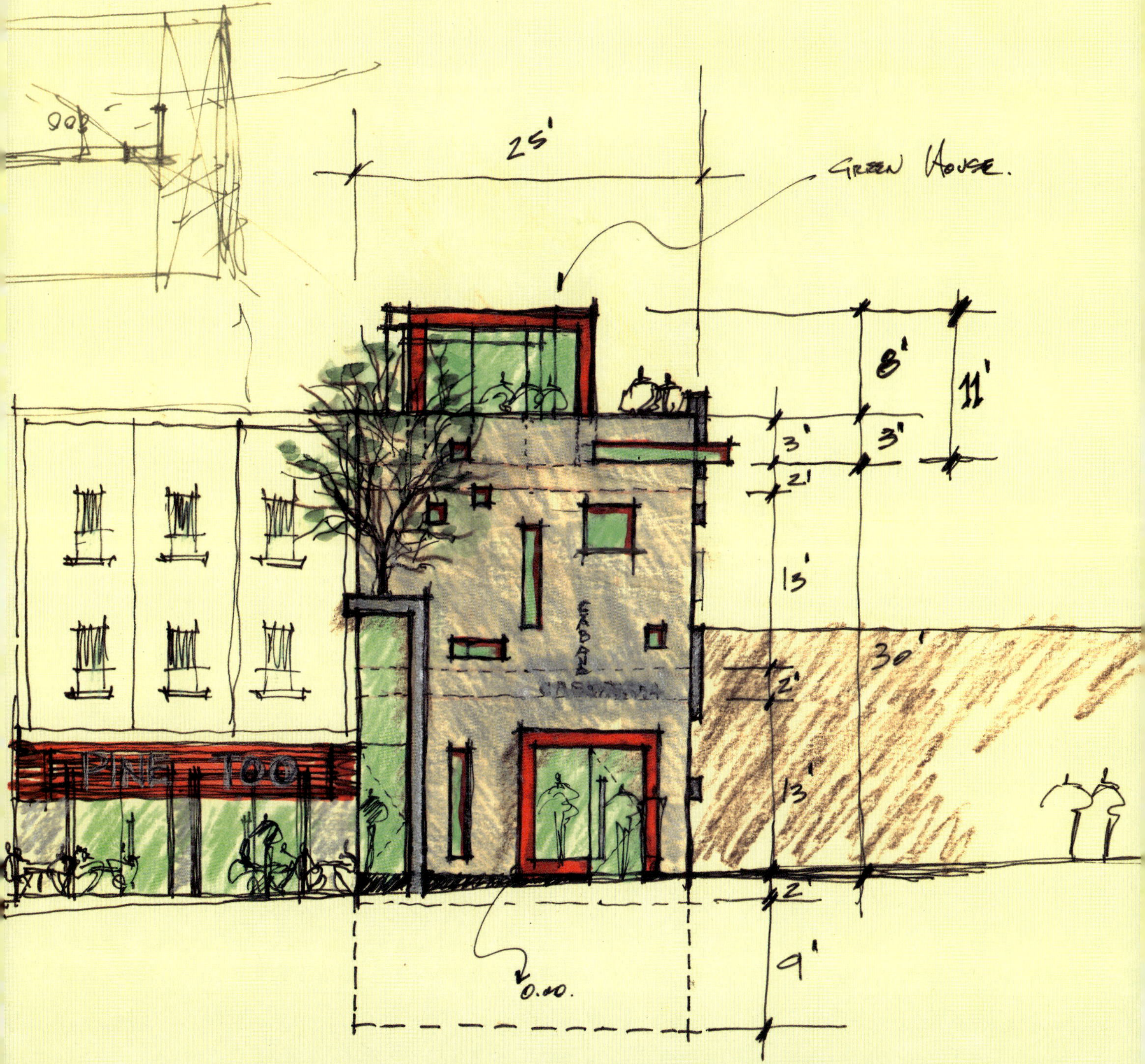

25'
GREEN HOUSE.
8'
11'
3'
3'
2'
13'
30'
2'
13'
PINE 100
2'
9'
0.00.

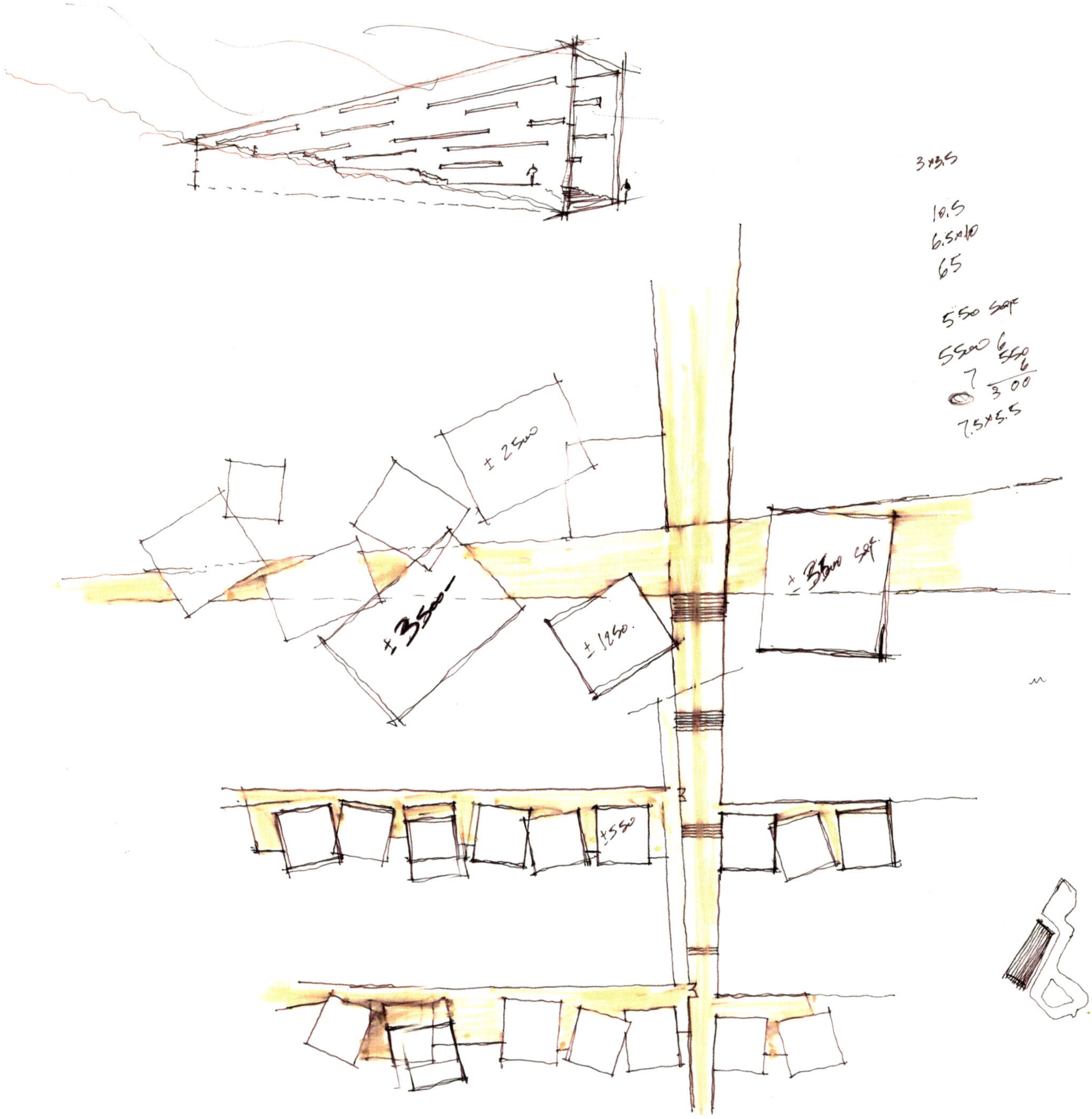
± 2500
± 3500
± 1850
± 550
7.5x5.5

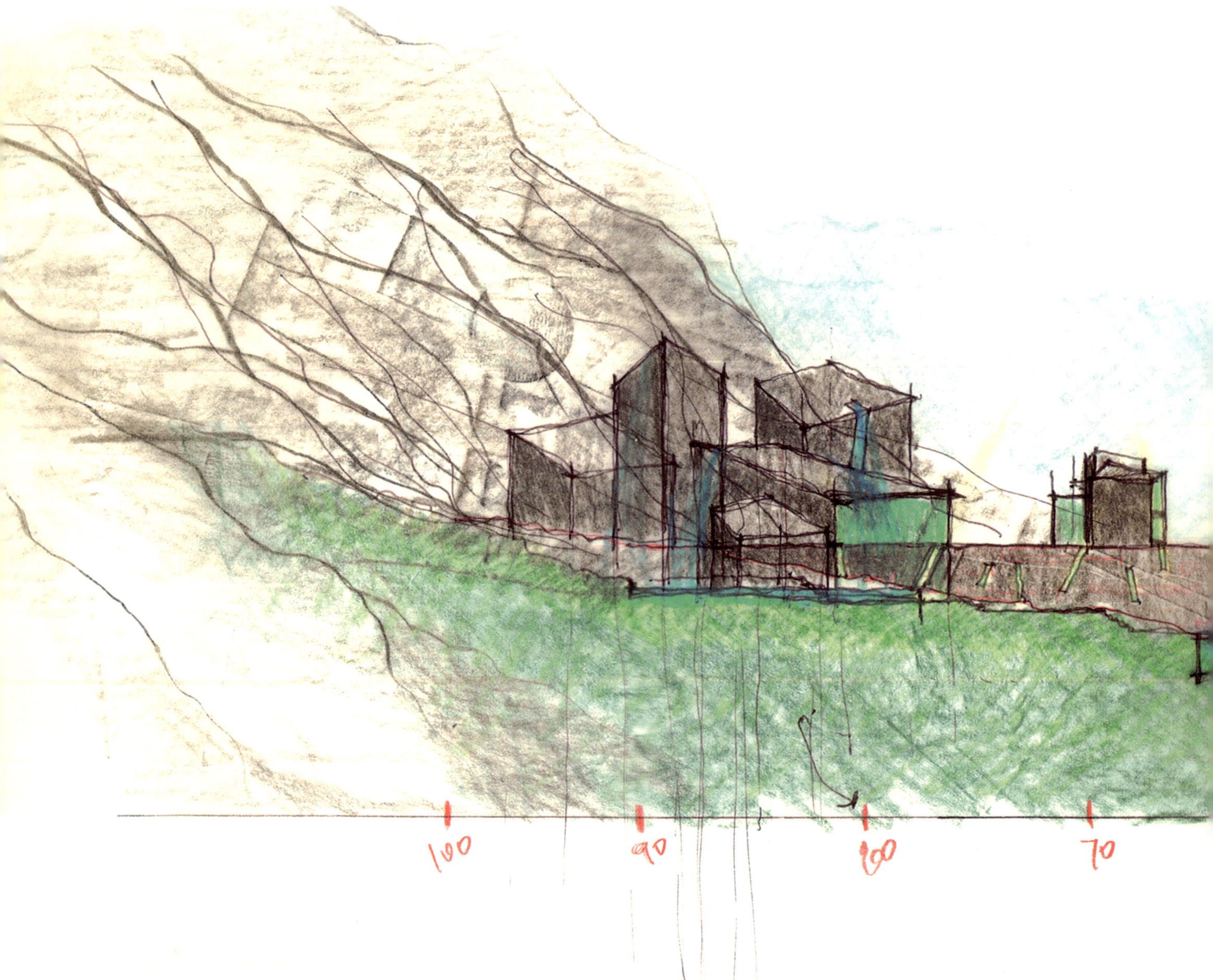
100
90
80
70

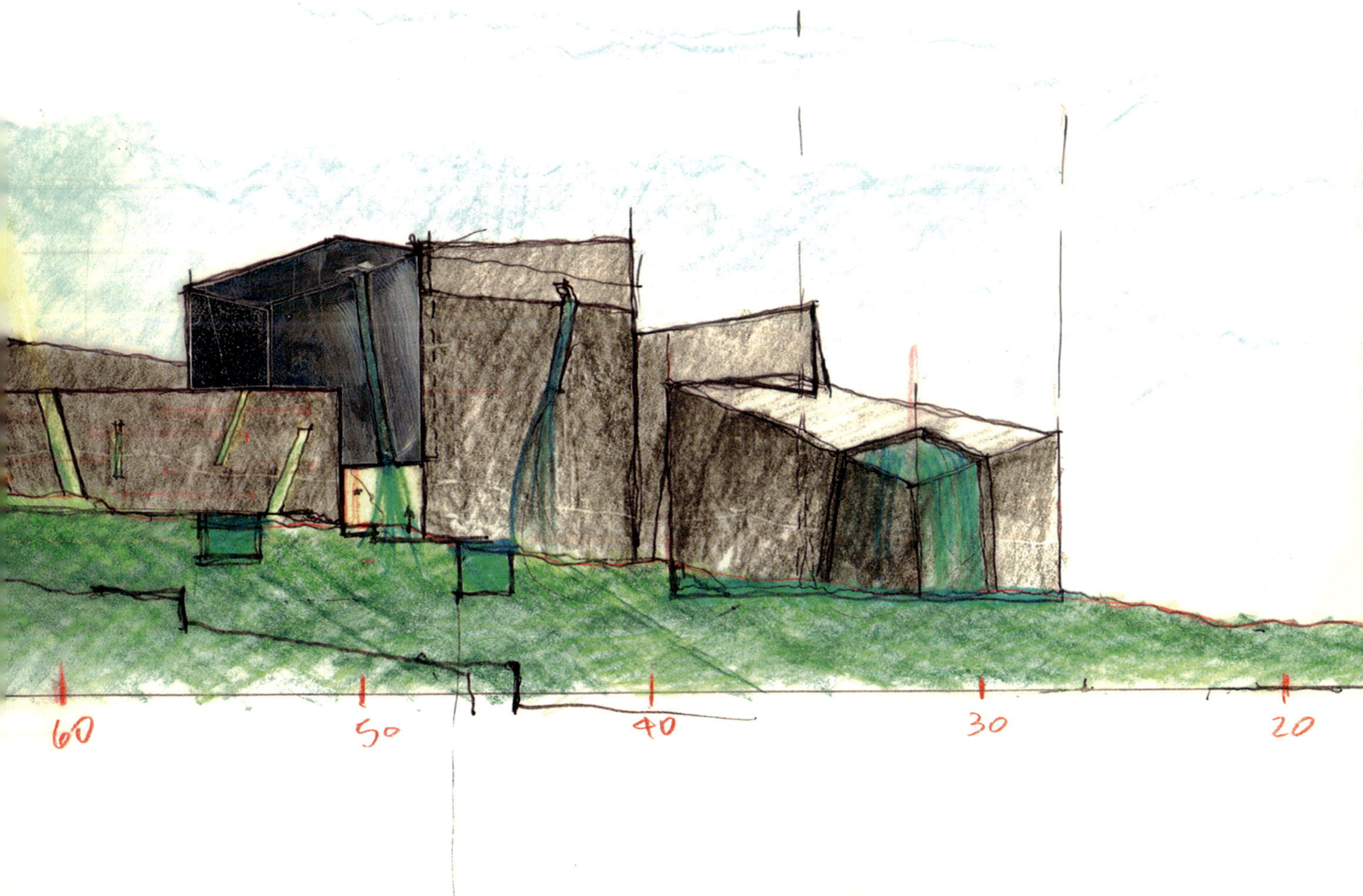
60
50
40
30
20

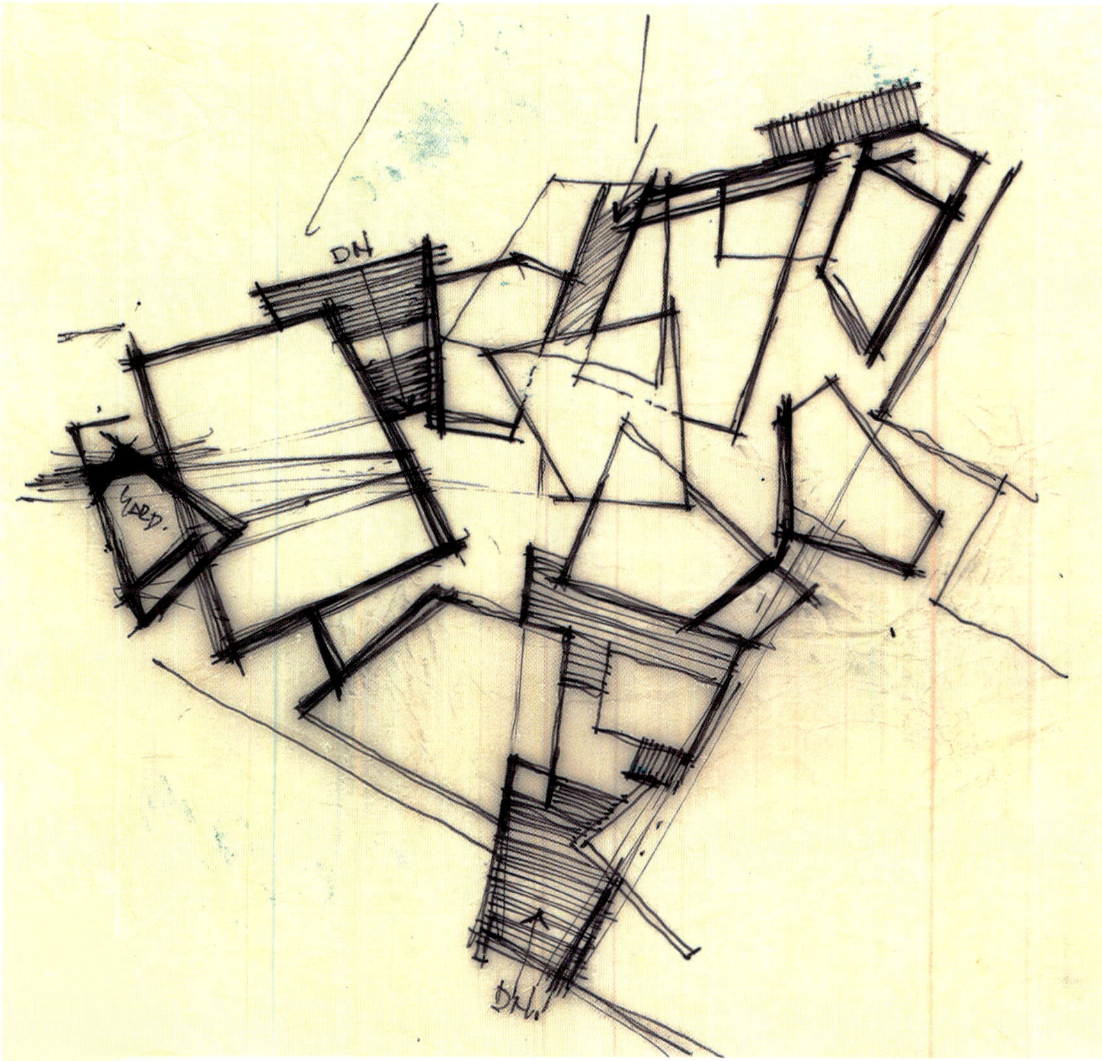
DN
YARD.
DN.

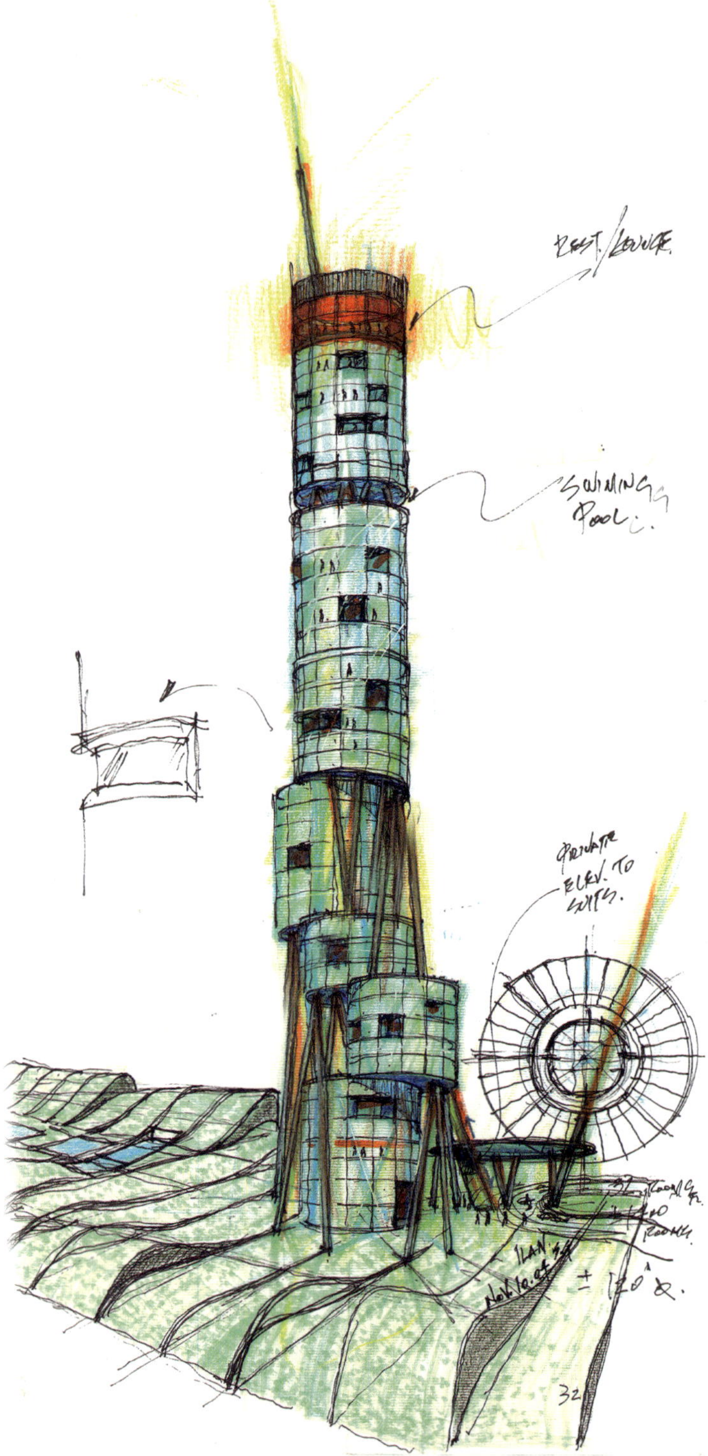
REST./LOUNGE
SWIMMING POOL
PRIVATE ELEV. TO SUITS.

LEVEL 1-7
LEVEL 8-12
LEVEL 13-15
LEVEL 16-21
LEVEL 22-38
LEVEL 39
LEVEL 40-45
LEVEL 46
LEVEL 48

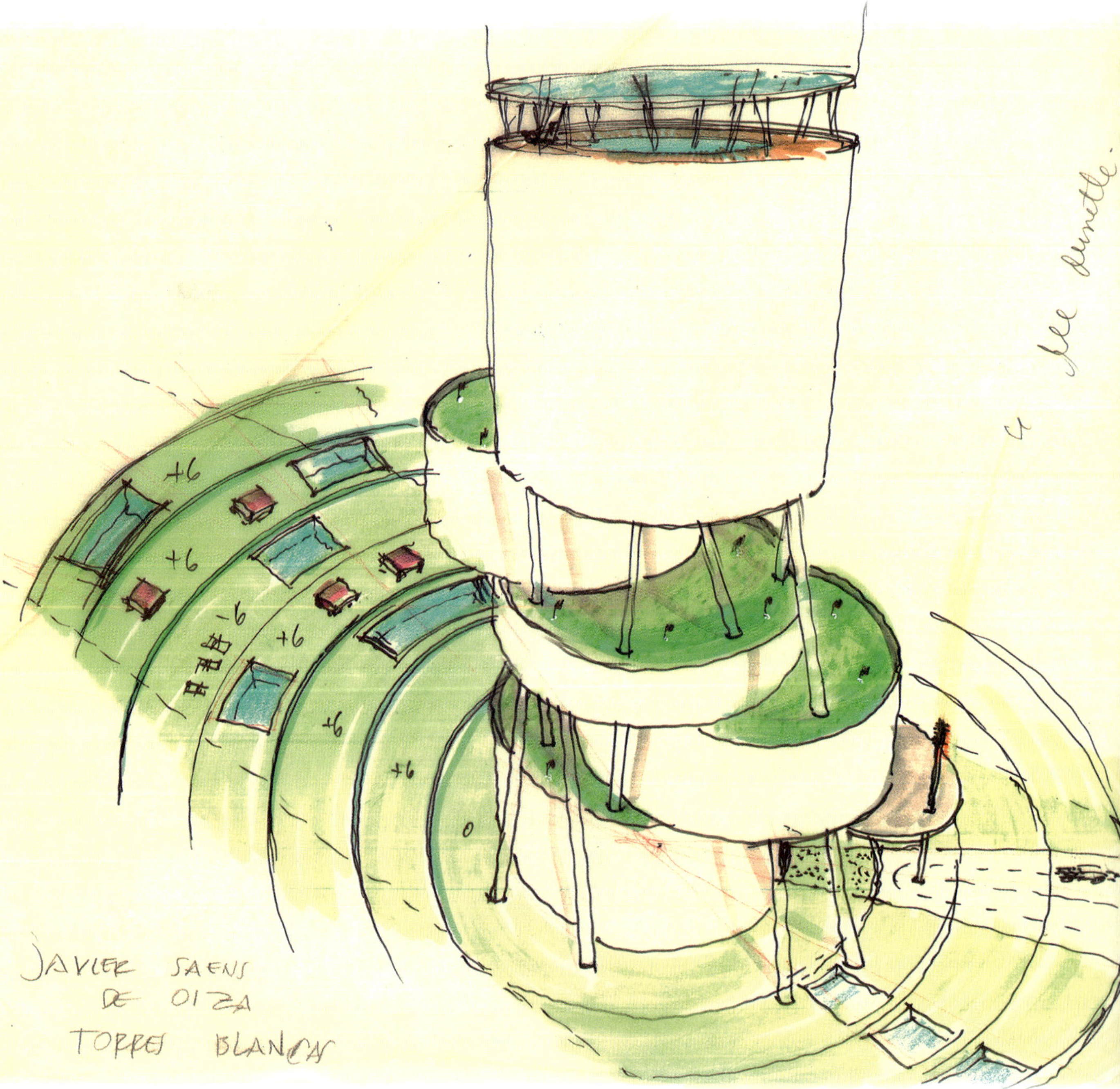
JAVIER SAENS
DE OIZA
TORRES BLANCAS

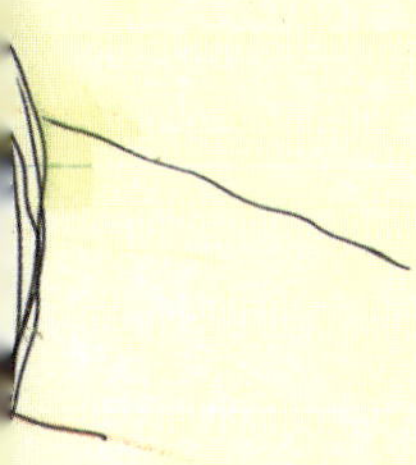

CLUB
CLUB.
30
10

CONRAD

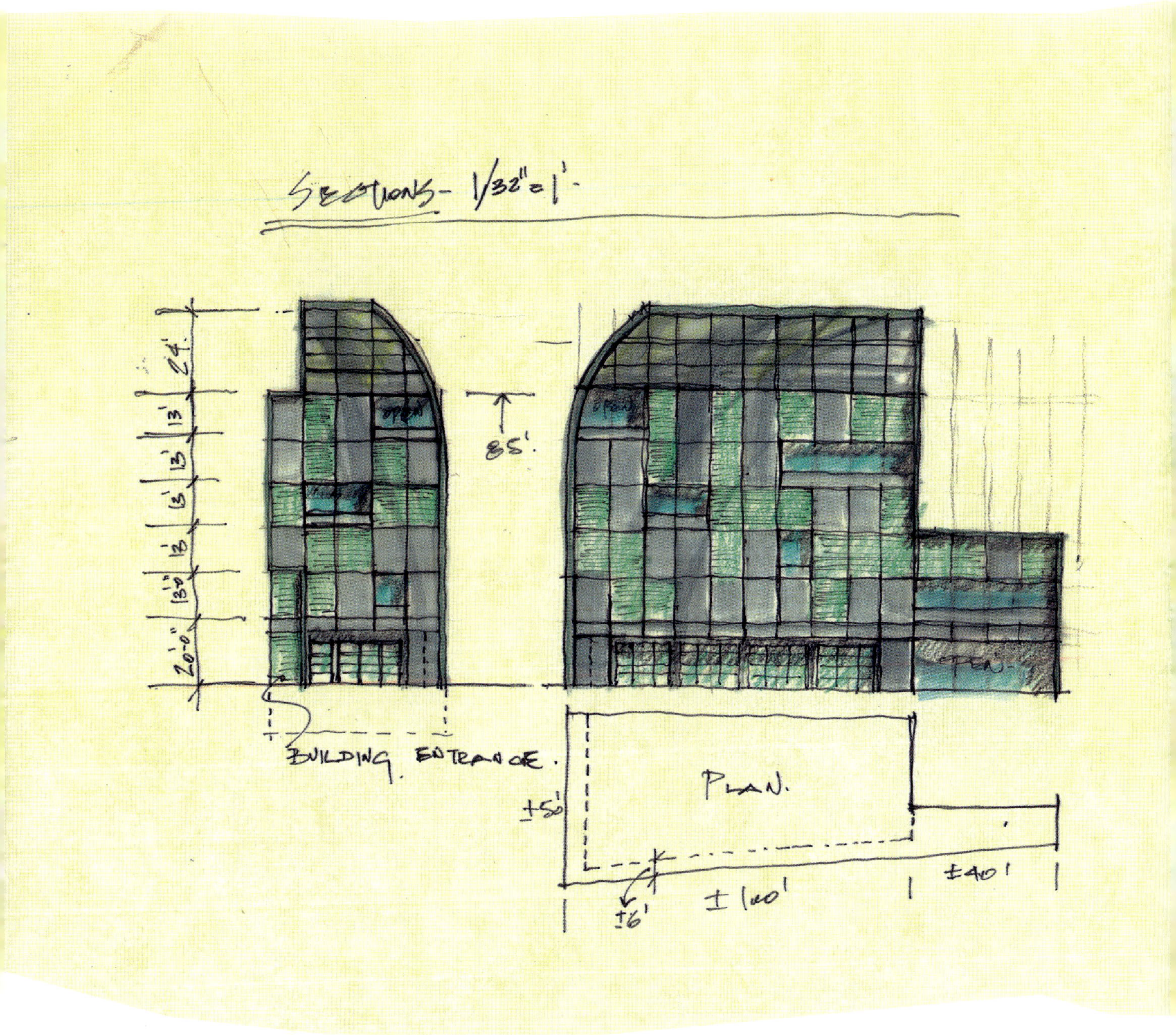
SECTIONS - 1/32" = 1'
24'
13'
13'
13'
13'
13'
20'-0"
OPEN
OPEN
85'
OPEN
BUILDING ENTRANCE.
PLAN.
±50'
±40'
±100'
±6'

ILAN 2007.

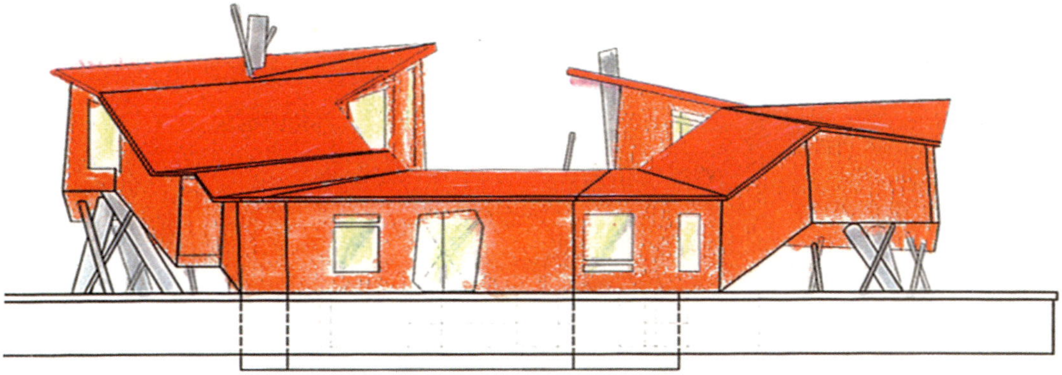

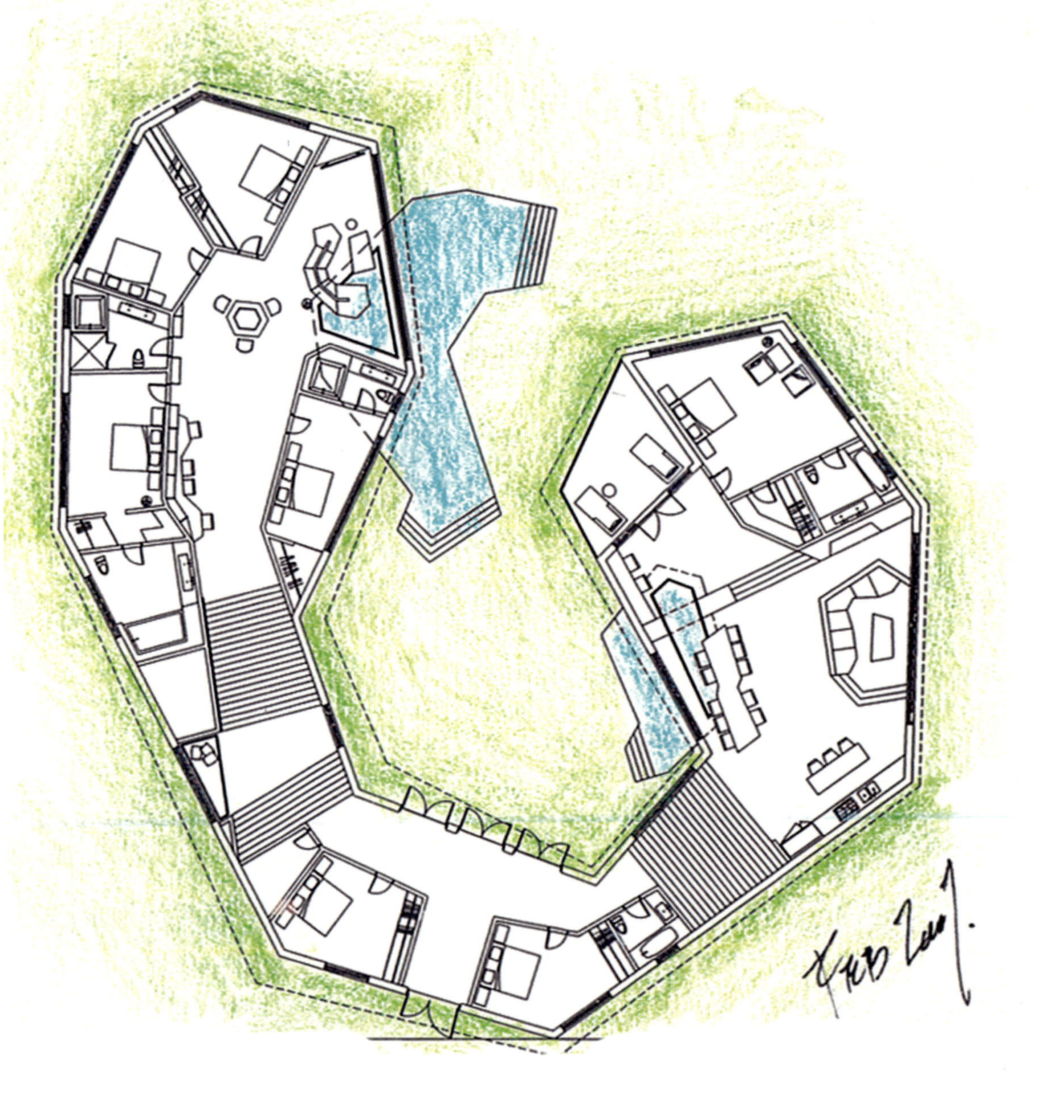

FEB. 2007

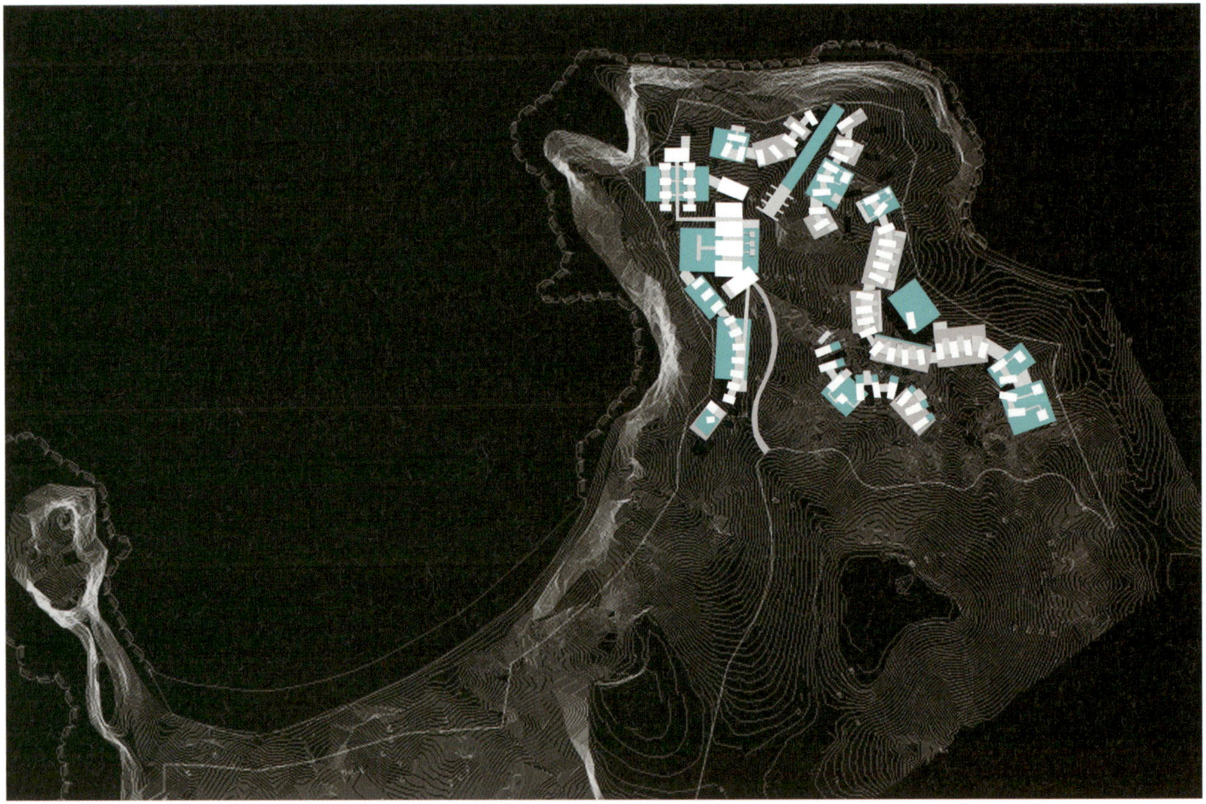

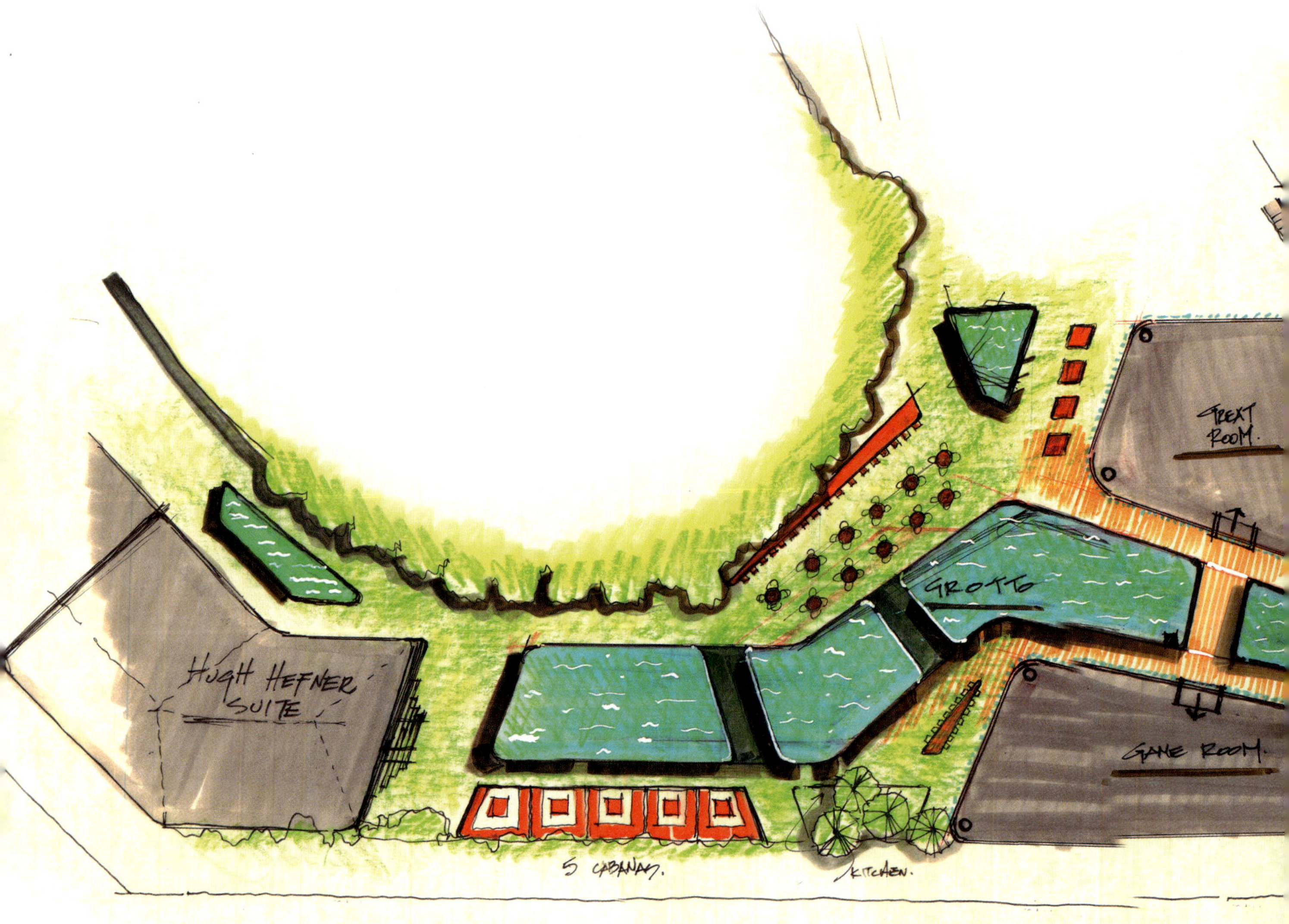
GREAT ROOM.
GROTTO
HUGH HEFNER SUITE
GAME ROOM.
5 CABANAS.
KITCHEN.

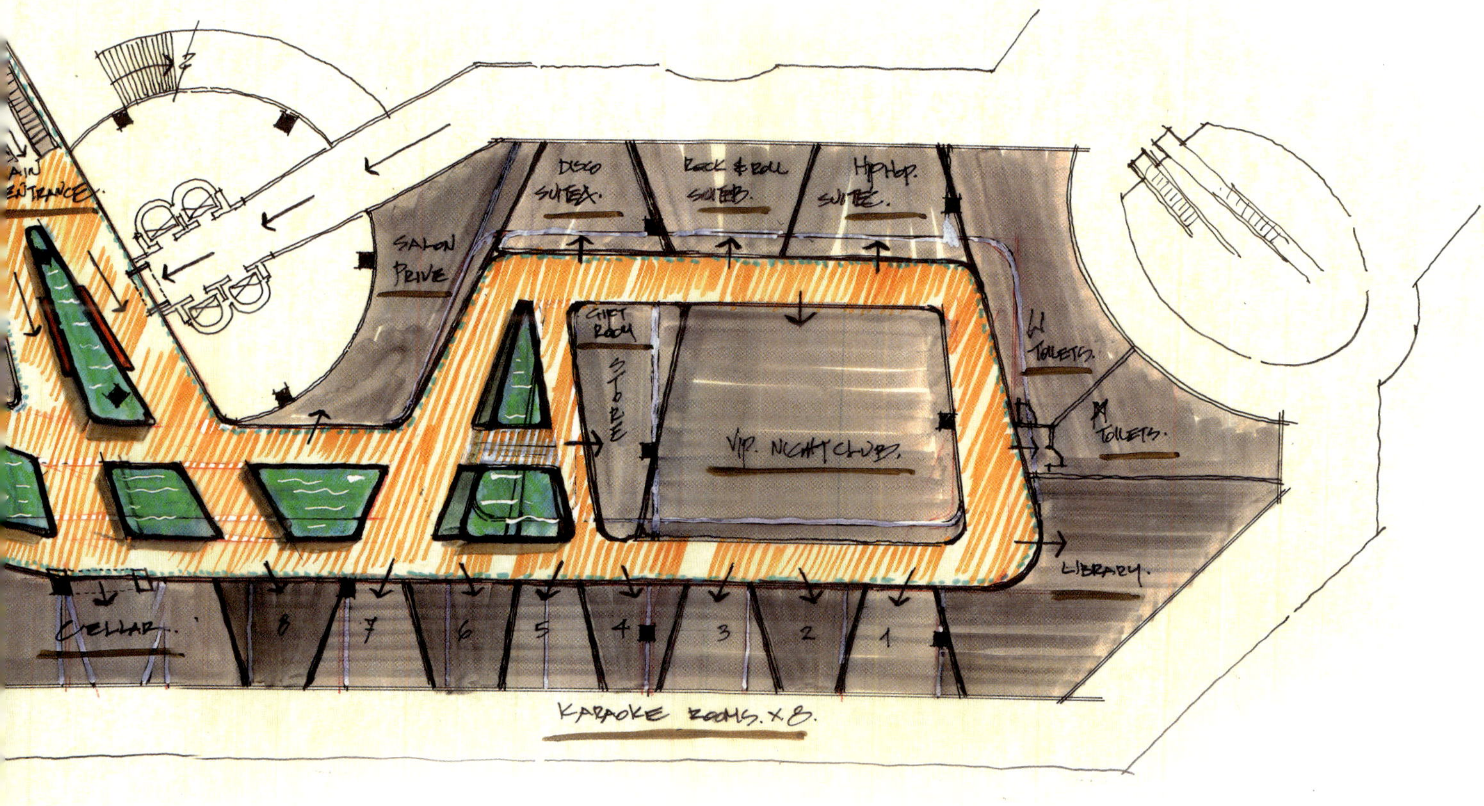

MAIN ENTRANCE.
SALON PRIVE
DISCO SUITE A.
ROCK & ROLL SUITE B.
HIP HOP. SUITE C.
GIFT ROOM
STORE
VIP. NIGHTCLUBS.
TOILETS.
TOILETS.
LIBRARY.
CELLAR.
8
7
6
5
4
3
2
1
KARAOKE ROOMS. X 8.

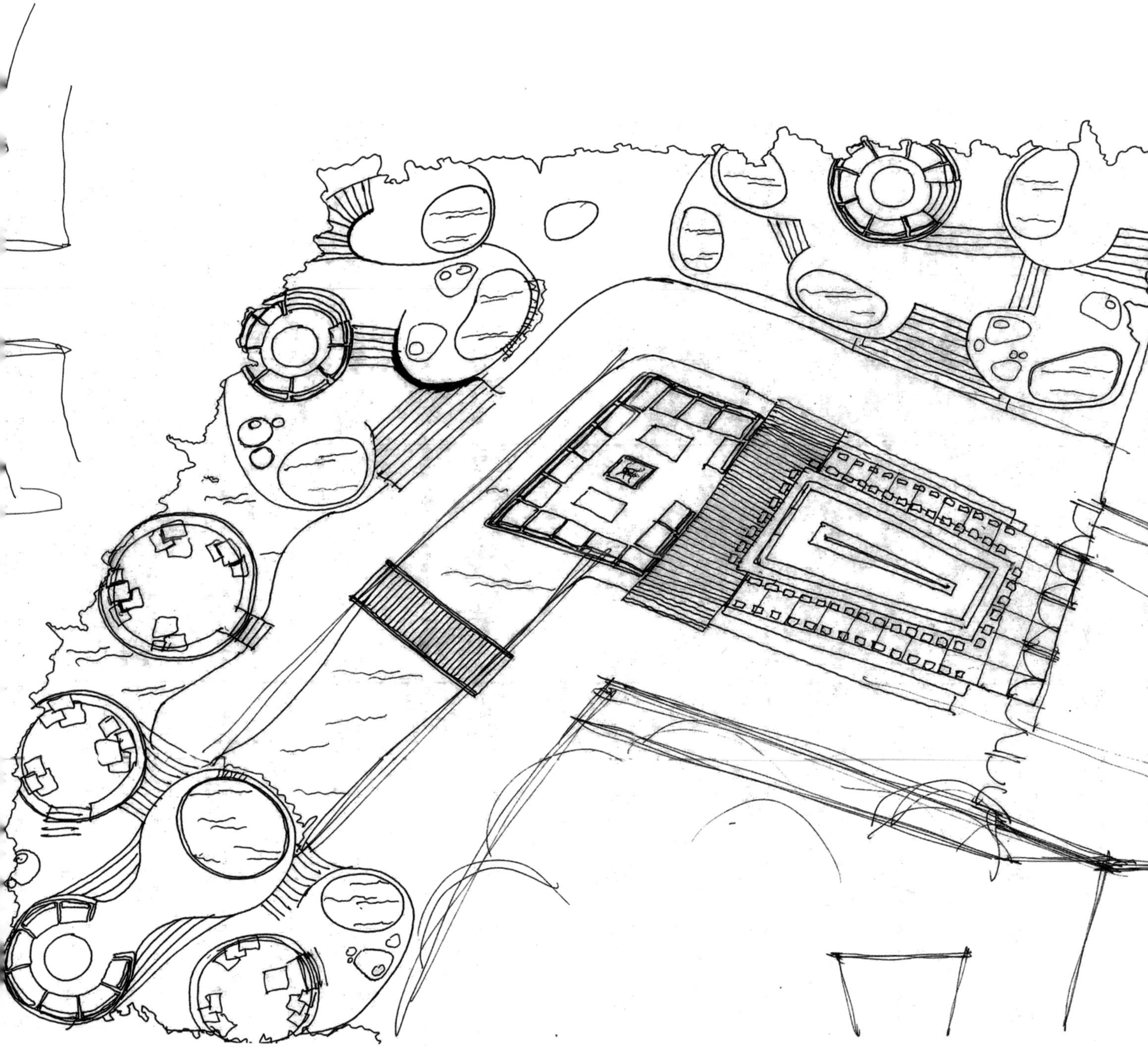

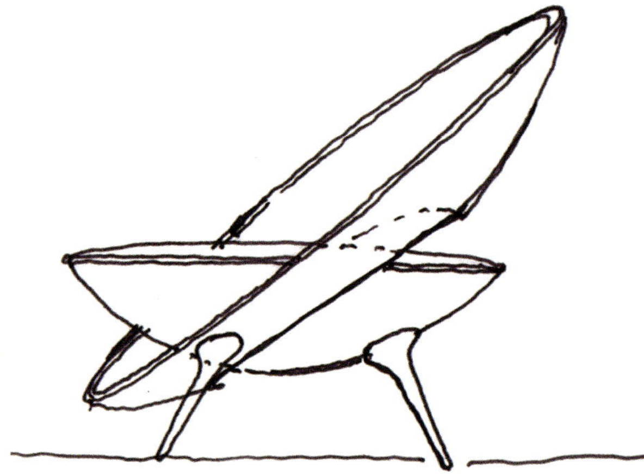

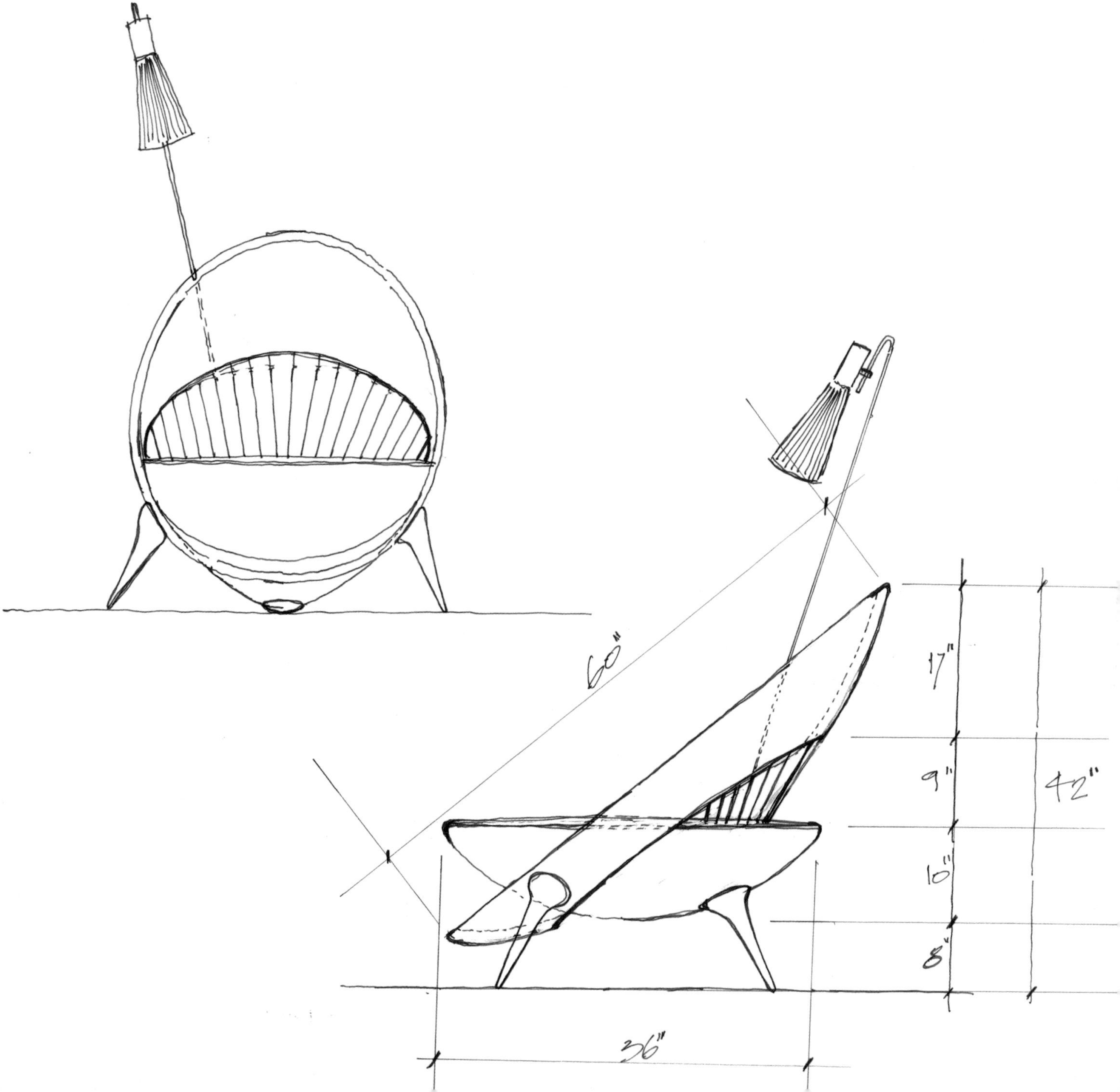

60"
17"
9"
42"
10"
8"
36"

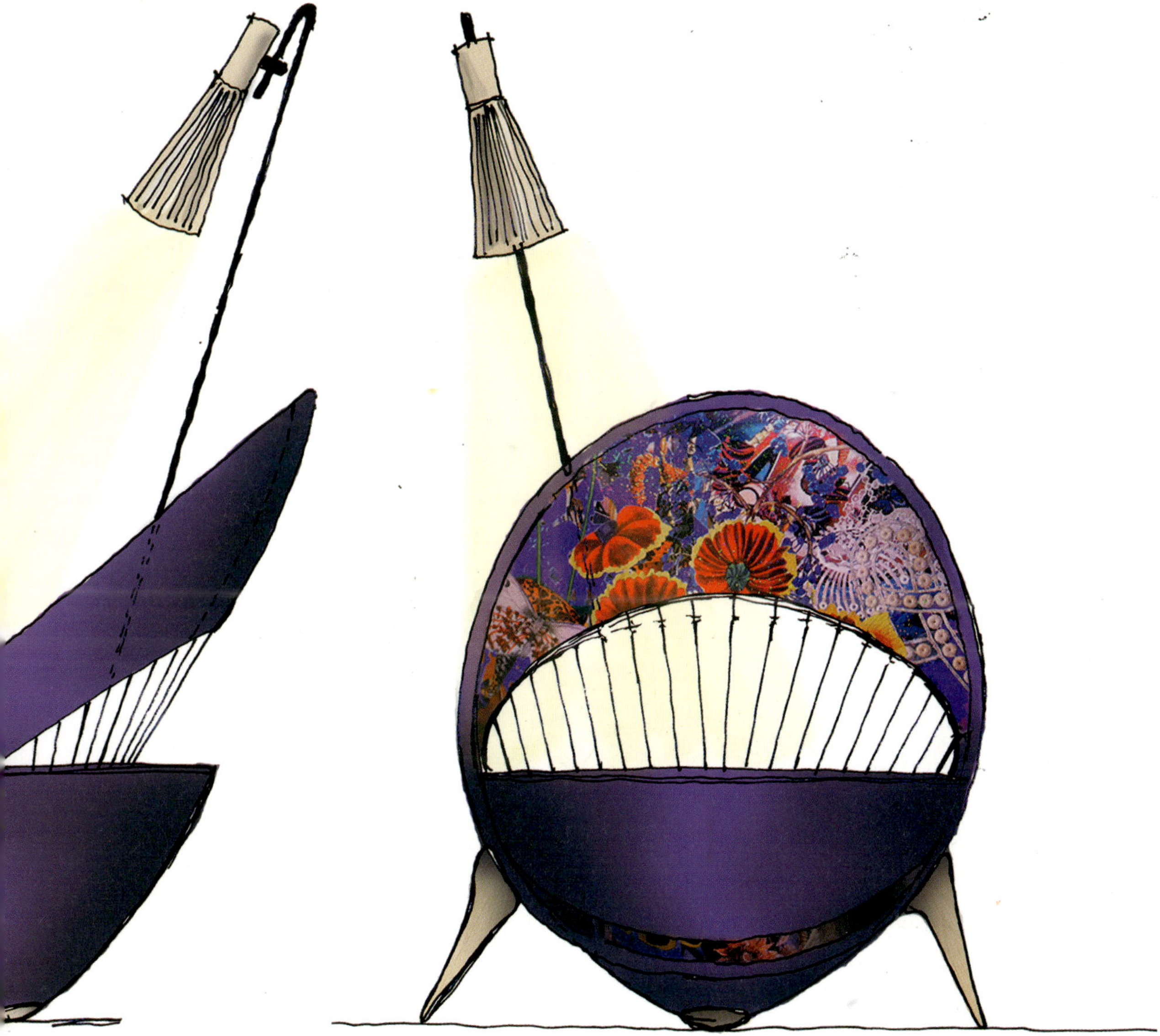

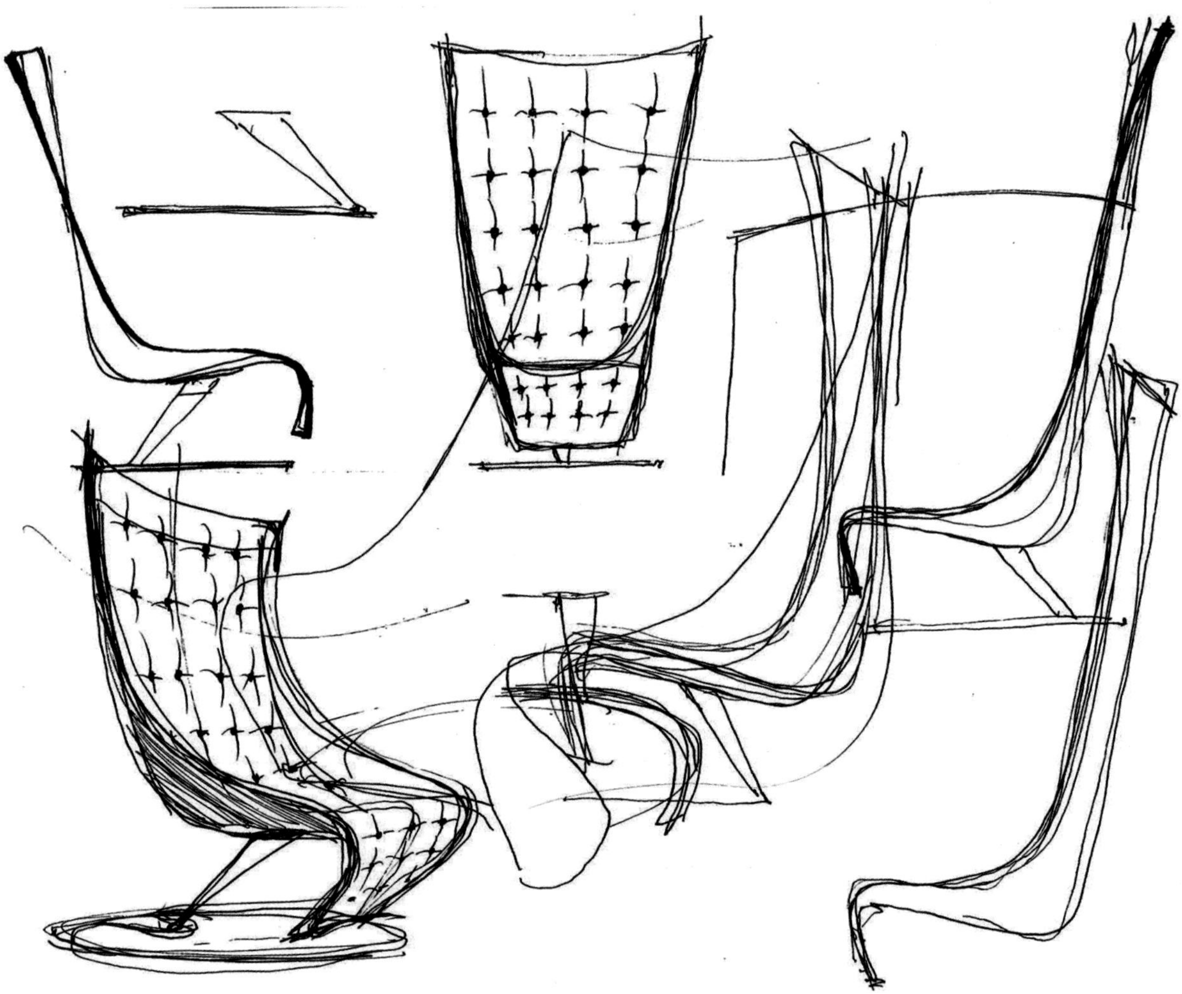

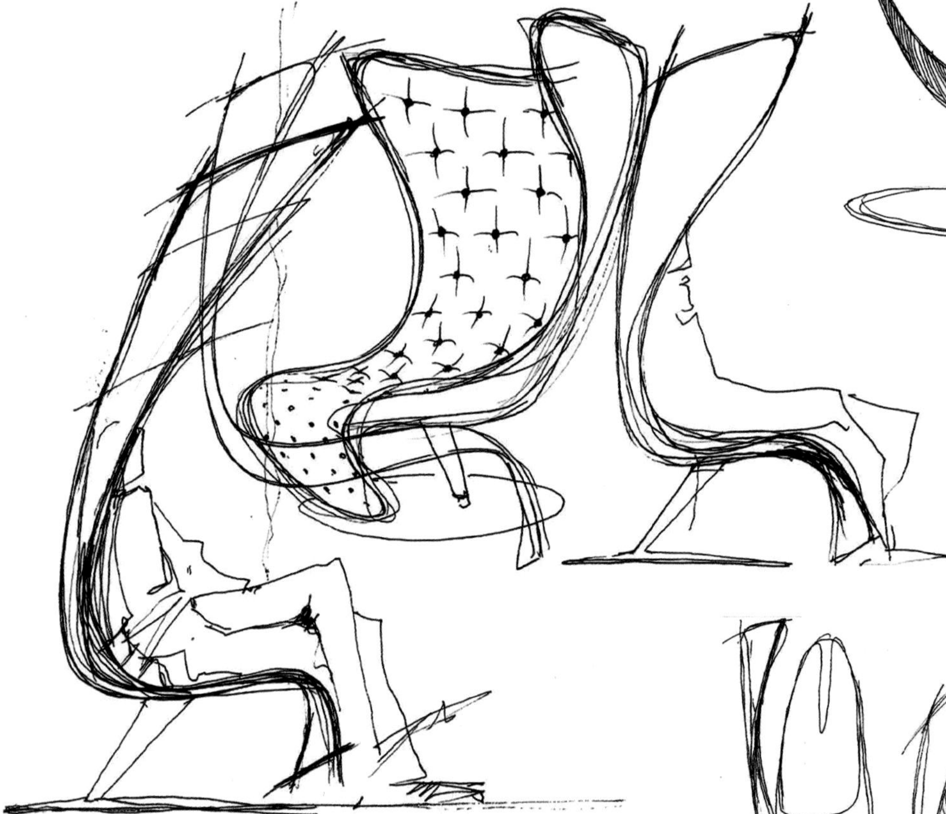

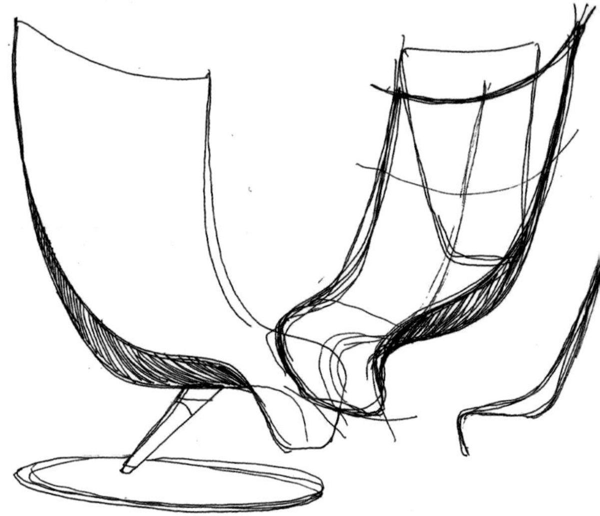

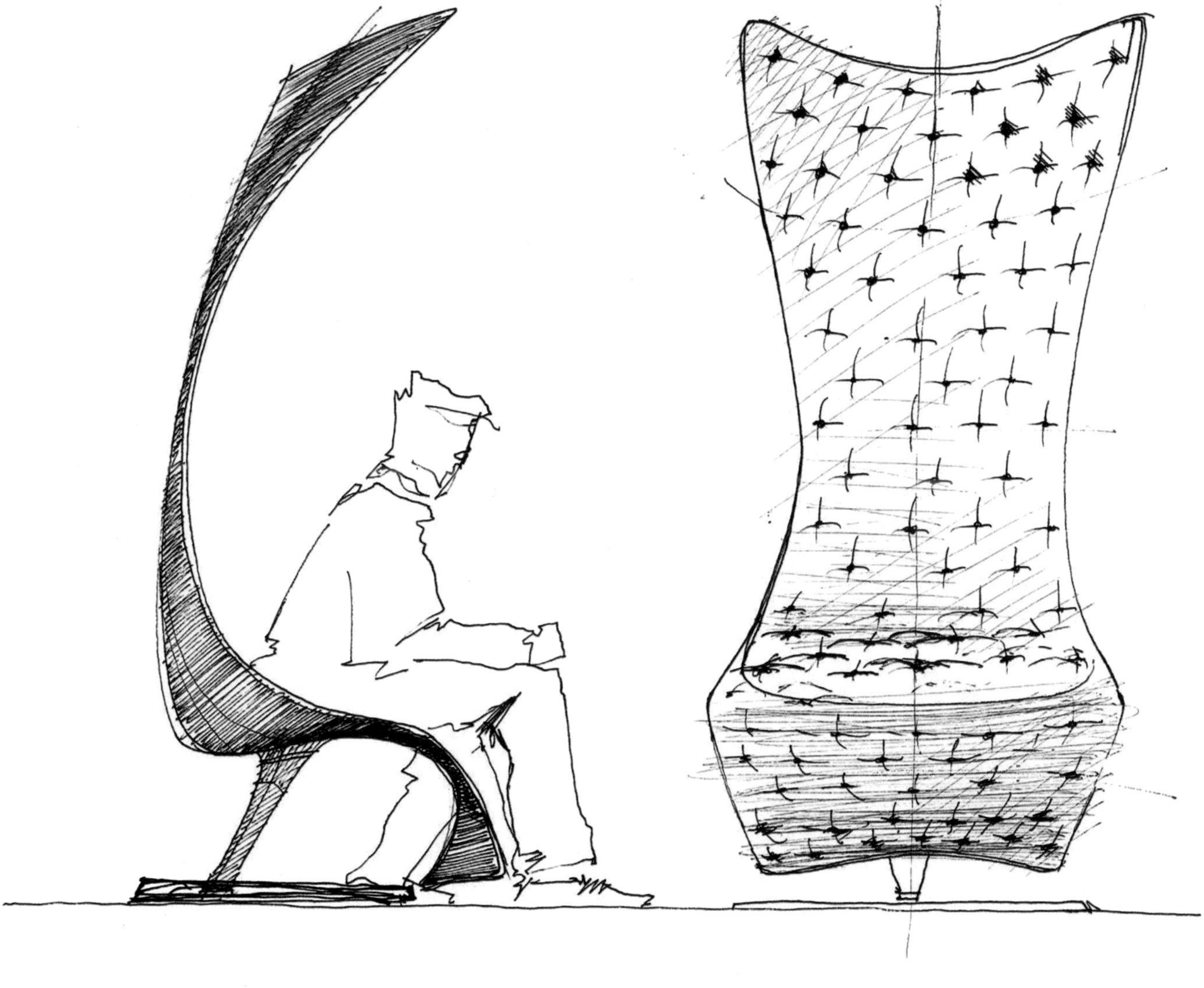

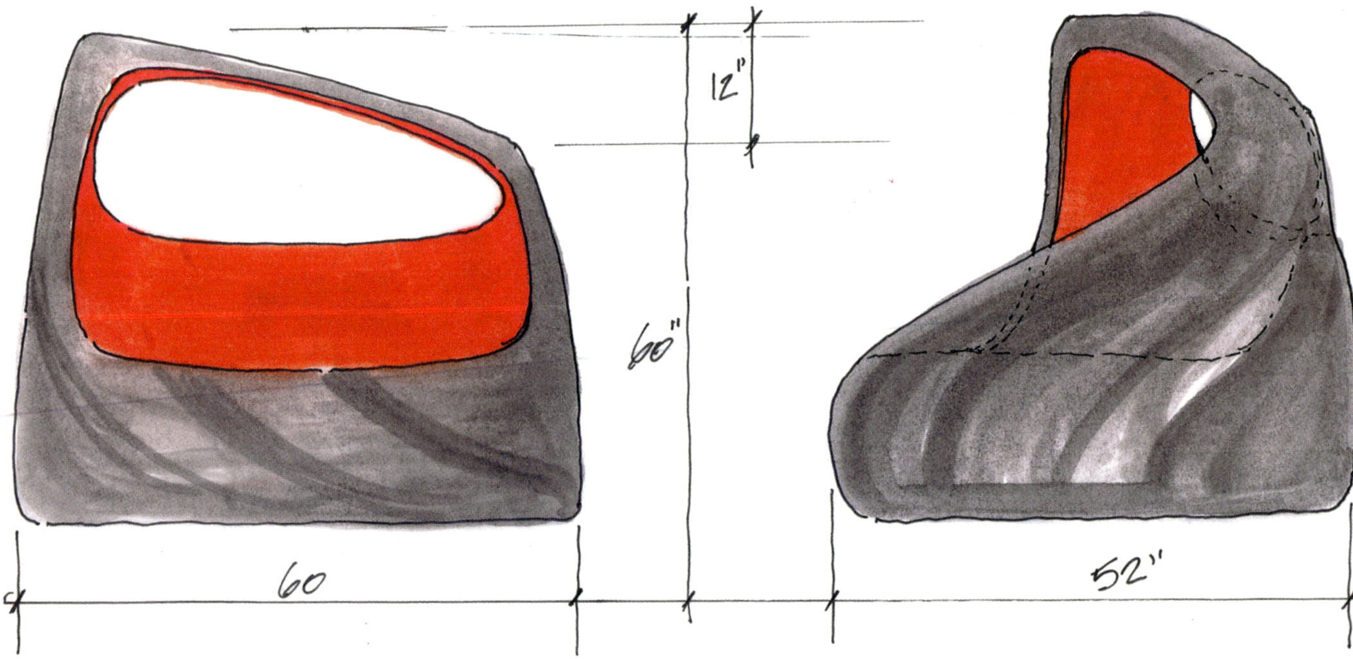
12"
60"
60
52"

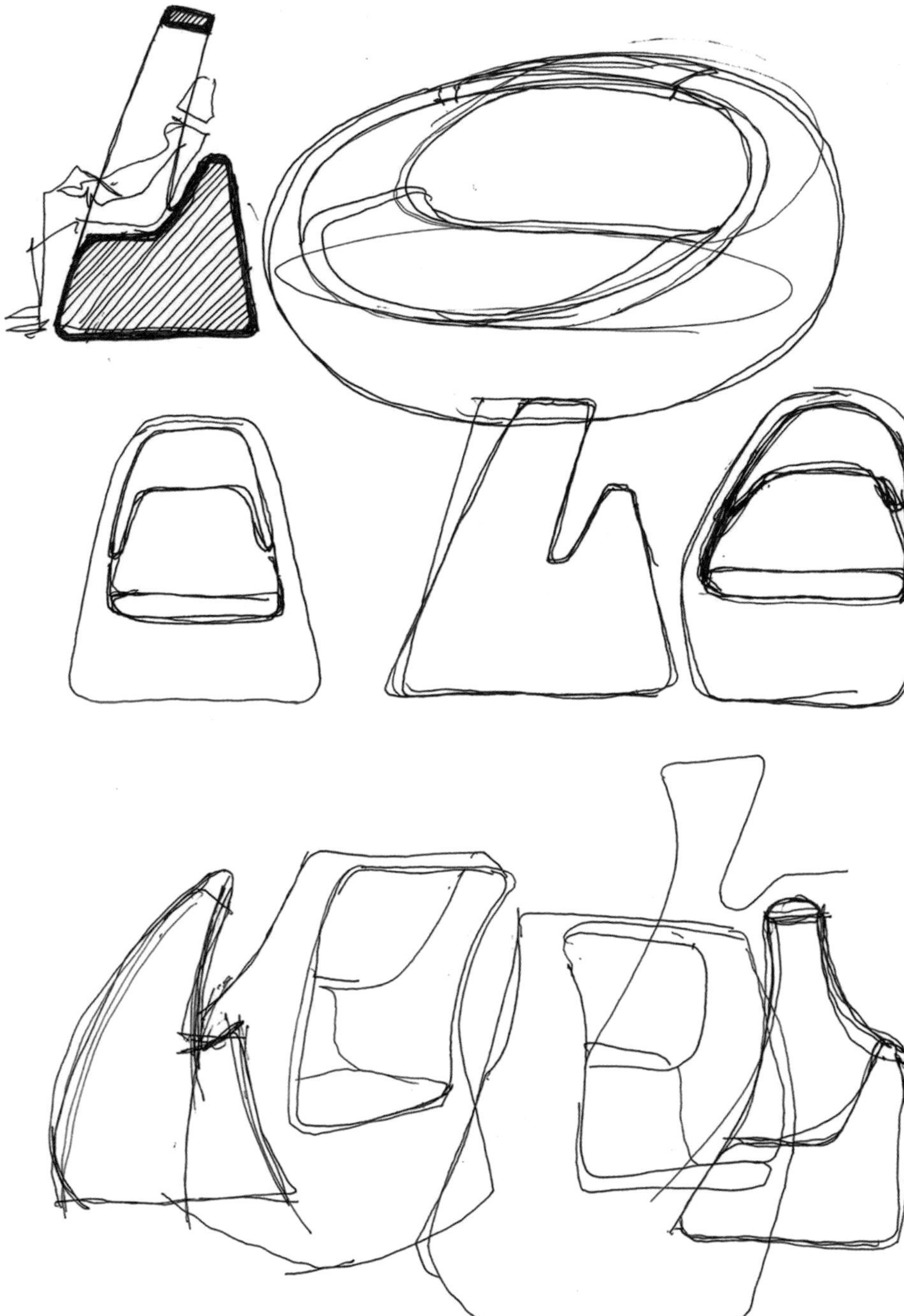

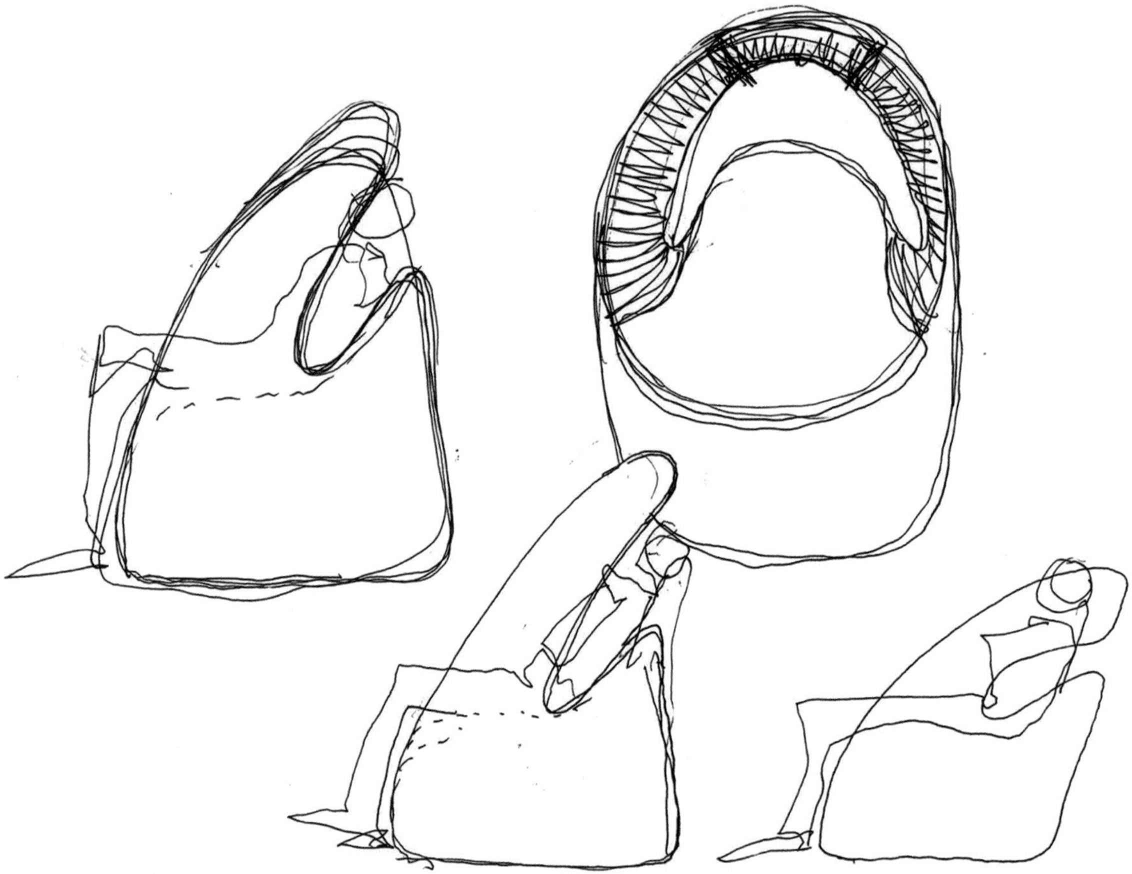

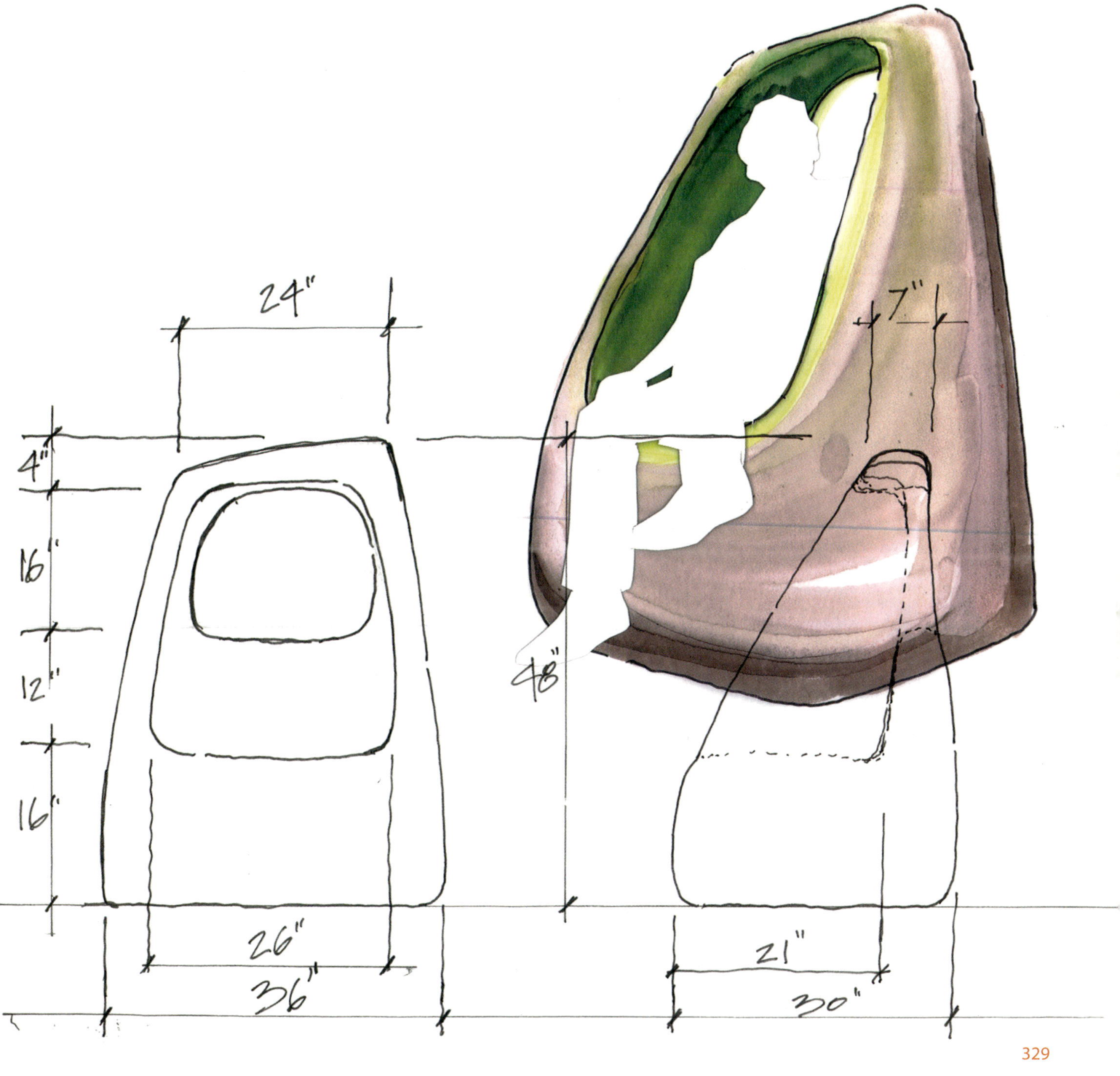
24"
7"
4"
16"
12"
16"
48"
26"
36"
21"
30"

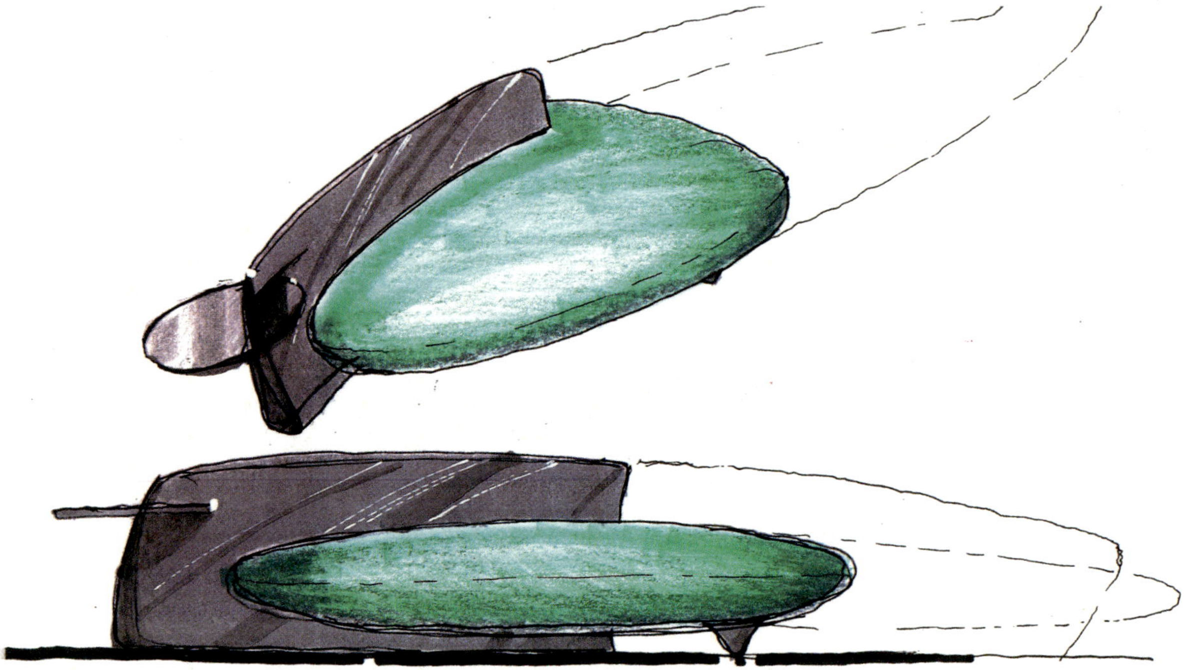

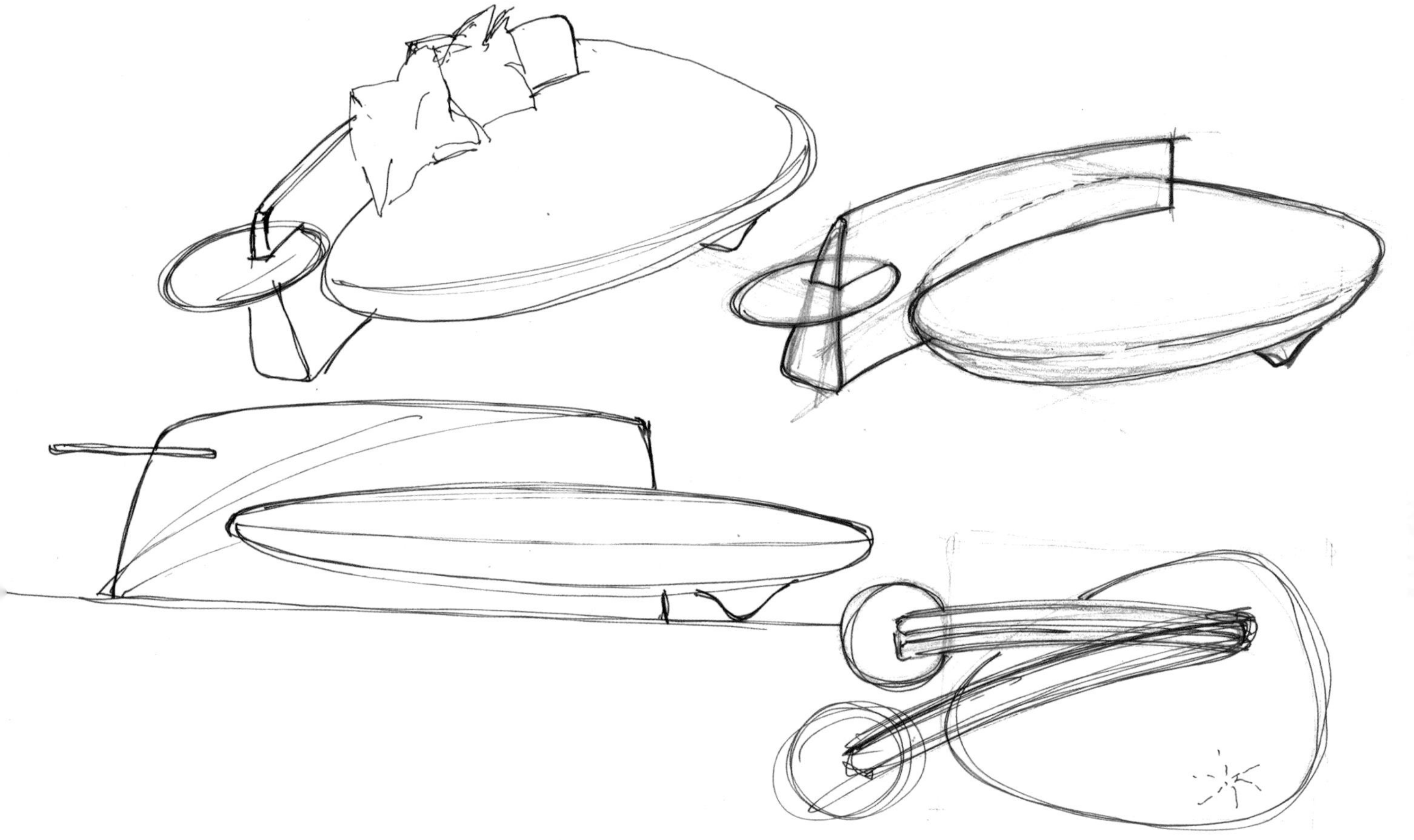

To my wife Ronit and my three doodles Gaia, Leeor, and Noam, the love of my life, for supporting me during the years and believing with me that imagination takes you everywhere.

RESTAURANTS & BARS

HOTELS

RESIDENTIALS

ARCHITECTURES

RETAILS

FURNITURES